Advances in

THE STUDY OF BEHAVIOR

VOLUME 5

Contributors to This Volume

CAROL DIAKOW

WALTER HEILIGENBERG

WILLIAM T. KEETON

D. J. McFARLAND

RONALD W. OPPENHEIM

KENNETH D. ROEDER

Advances in
THE STUDY OF BEHAVIOR

Edited by

Daniel S. Lehrman
Institute of Animal Behavior
Rutgers University
Newark, New Jersey

Jay S. Rosenblatt
Institute of Animal Behavior
Rutgers University
Newark, New Jersey

Robert A. Hinde
Medical Research Council
Unit on the Development and Integration of Behaviour
University Sub-Department of Animal Behaviour
Madingley, Cambridge, England

Evelyn Shaw
Department of Biological Sciences
Stanford University
Stanford, California

──────────── **VOLUME 5** ────────────

ACADEMIC PRESS   New York   San Francisco   London   1974
A Subsidiary of Harcourt Brace Jovanovich, Publishers

ACADEMIC PRESS, INC.
111 Fifth Avenue, New York, New York 10003

United Kingdom Edition published by
ACADEMIC PRESS, INC. (LONDON) LTD.
24/28 Oval Road, London NW1

LIBRARY OF CONGRESS CATALOG CARD NUMBER: 64-8031

ISBN 0–12–004505–2

PRINTED IN THE UNITED STATES OF AMERICA

Contents

Some Neuronal Mechanisms of Simple Behavior
KENNETH D. ROEDER

The Orientational and Navigational Basis of Homing in Birds
WILLIAM T. KEETON

The Ontogeny of Behavior in the Chick Embryo
RONALD W. OPPENHEIM

Processes Governing Behavioral States of Readiness
WALTER HEILIGENBERG

Time-Sharing as a Behavioral Phenomenon
D. J. McFARLAND

Male-Female Interactions and the Organization of
Mammalian Mating Patterns
CAROL DIAKOW

List of Contributors

Numbers in parentheses indicate the pages on which the authors' contributions begin.

CAROL DIAKOW,* *The Rockefeller University, New York, New York* (227)

WALTER HEILIGENBERG,† *Max-Planck-Institut für Verhaltensphysiologie, Seewiesen, Germany* (173)

WILLIAM T. KEETON, *Section of Neurobiology and Behavior, Cornell University, Ithaca, New York* (47)

D. J. McFARLAND, *Department of Experimental Psychology, Oxford University, Oxford, England* (201)

RONALD W. OPPENHEIM, *Neuroembryology Laboratory, Division of Research, North Carolina Department of Mental Health, Raleigh, North Carolina* (133)

KENNETH D. ROEDER, *Department of Biology, Tufts University, Medford, Massachusetts* (1)

* Present Address: Department of Biology, Adelphi University, Garden City, Long Island, New York.

† Present Address: Scripps Institution of Oceanography, University of California at San Diego, La Jolla, California.

Preface

Shortly before the death of Daniel S. Lehrman, the Editors of *Advances in the Study of Behavior* welcomed Jay S. Rosenblatt to the staff. For several months he worked alongside the Editor. He has now been asked to serve as Chief Editor of the series. Colin G. Beer has been invited to serve as the fourth member of the editorial board starting with Volume 6. The Editors will continue to be guided by the policy statement which appeared in the preface to the first volume.

"The study of animal behavior is attracting the attention of ever-increasing numbers of zoologists and comparative psychologists in all parts of the world, and is becoming increasingly important to students of human behavior in the psychiatric, psychological, and allied professions. Widening circles of workers, from a variety of backgrounds, carry out descriptive and experimental studies of behavior under natural conditions, laboratory studies of the organization of behavior, analyses of neural and hormonal mechanisms of behavior, and studies of the development, genetics, and evolution of behavior, using both animal and human subjects. The aim of *Advances in the Study of Behavior* is to provide workers on all aspects of behavior an opportunity to present an account of recent progress in their particular fields for the benefit of other students of behavior. It is our intention to encourage a variety of critical reviews, including intensive factual reviews of recent work, reformulations of persistent problems, and historical and theoretical essays, all oriented toward the facilitation of current and future progress. *Advances in the Study of Behavior* is offered as a contribution to the development of cooperation and communication among scientists in our field."

Contents of Previous Volumes

DANIEL S. LEHRMAN
1919–1972

Dedication
DANIEL S. LEHRMAN

This volume was in preparation when the Chief Editor of this serial publication, Daniel S. Lehrman, died in August, 1972. Editing *Advances in the Study of Behavior* was natural for Dan since it gave him and co-editors Robert A. Hinde and Evelyn Shaw the opportunity to present varied viewpoints in the field of animal behavior, to encourage researchers whose work was not yet well known, to enable the elders in our field to write fully of their life's work and thinking, and to spotlight new developments in theory and research in animal behavior. Dan's interest in animal behavior became deeply rooted in him during his adolescence; animal behavior had an excitement for him that was not based upon theory and could not be constrained by the discipline of scientific method. It was, as he wrote, an expression of his relationship to the natural world around him: understanding animal behavior was for him a means by which he grew to understand himself. This volume is dedicated to his work and his life by his co-editors, and by the publishers of this series.

Some Neuronal Mechanisms of Simple Behavior

KENNETH D. ROEDER[1]

DEPARTMENT OF BIOLOGY
TUFTS UNIVERSITY
MEDFORD, MASSACHUSETTS

[1] Much of the personal research reported in this paper was supported by research grants and a Research Career Award from the National Institutes of Health.

1

I. INTRODUCTION

At the present time there is considerable interest in the mechanisms of animal behavior—the means whereby animals do what they do. The problem is the direct or indirect concern of a variety of fields ranging from the molecular to the ecological levels of zoology. However, the complex technical requirements of many of these disciplines tend to obscure the relevance of their specific researches in such a broad context as this. In other words, the increasing sophistication of scientific technology makes it ever harder to remain a generalist and become at the same time a specialist.

The fields of ethology and neurophysiology lie not far apart on this "spectrum" of disciplines. They are subject to the general tendency toward specialization so that ethologists and neurophysiologists find most of their energies directed to parochial concerns. Yet, closer mutual understanding would seem desirable since both deal directly with animal behavior, ethology with its origins and causes in intact animals and neurophysiology with its mechanisms. This essay seeks points of interchange where information and ideas might flow more freely between them.

This broad objective forces me into another form of parochialism. I shall make no attempt to examine more than a fraction of the situations where ethology and neurophysiology may eventually find congruence. Instead I shall compare the possible origins of both fields as a source of their basic viewpoints and attempt to present some of the concepts of neurophysiology that seem relevant to behavior. I shall then examine some of the misapprehensions of each about the other and search for points of contact in connection with problems within my own purview. In short, this is in no sense a comprehensive survey but rather a very personal view of the matter.

II. ORIGINS

The nature of the relation between ethology and neurophysiology may be appreciated by considering their origins and how these may have influenced their present attitudes. By far the greatest influence that transformed the early folklore of animal behavior into ethology as it is today came undoubtedly from the principles of evolution and species formation first perceived by Charles Darwin. Students of animal behavior found that they were studying "alive and on-stage," as it were, the current action of evolution. Taxonomic affinities and the theory of natural selection provided them with frames of reference in which to organize their observations. Students of animal behavior were slower than comparative anatomists to realize this opportunity even though they were closer to the actual performance of natural selection. This was probably because structures can be studied and stored in museum jars whereas behavior patterns cannot.

Modern ethology dates roughly from Heinroth and Whitman and received the main impetus for its development since World War II from the demonstrations by Konrad Lorenz and Niko Tinbergen and their students that certain behavior patterns are just as characteristic of species as is their form. In addition, concepts developed by this group such as innate releasing mechanisms, displacement activities, intention movements, sign stimuli, imprinting, and so on provided an immense intellectual stimulus to the study of animal behavior. Even though many of these concepts have since been transmuted they served to attract many agile minds to the field by providing hints of a general order and logic in the behavior of animals. A central guidepost of ethology can be said to be the causation of behavior and its adaptedness and survival value for breeding populations; its method is to observe with minimum disturbance of the natural situation and to intervene experimentally only with great caution.

Neurophysiology, considered as a branch of physiology, probably developed as a practical lore, being associated closely with medicine during historic times. Since physiology became concerned with analyzing mechanisms quite early in its history it had the advantage of cross-fertilization with chemistry, physics, and pharmacology, and today with molecular biology. These provided physiology with a conceptual framework and a linkage with general science that were denied the study of animal behavior until Darwin's time. During the present century the framework of neurophysiology has been further strengthened by the work of men such as Sherrington, Gasser and Erlanger, Hill, Adrian, Hodgkin and Huxley, and Eccles, to mention only a few names. Its scope has been greatly enlarged by advances at the molecular and biochemical levels, by tools such as electron microscopy, and by vast technical improvements in the methods of electronic amplification, signal registration, and data analysis.

In addition to its concern with the workings of the human nervous system neurophysiology has become increasingly comparative. Interest in the nervous systems of invertebrates has been active in Europe for the past century. It was early recognized that an understanding of the central nervous system as prime mover of behavior depended on comprehending transactions in a complex meshwork of neurons, and that the meshwork of invertebrate nervous systems contains many thousands of times fewer neurons than that of higher vertebrates. Many species were sought out for experimentation, and neurophysiology found the need to take stock of evolutionary relationships—the "backbone" of ethology. Details of this development are to be found in Prosser and Brown (1961) and Bullock and Horridge (1965).

Thus, their separate histories endowed ethology and neurophysiology with different attitudes toward animal behavior. These attitudes may be summarized as follows. Neurophysiology is primarily concerned with the action of the nervous system "inside the skin"; it is mainly analytical in inclination which carries it to molecular levels; it deals with mechanisms, that is, with questions such as "How does it work?" Ethology is concerned with the actions of intact animals, its purview being "outside the skin"; it is synthetic in inclination, an important part of its work being to assemble ethograms or inventories of the behavior characteristic of species; and it deals with the causation of behavior and with adaptedness, that is, with questions such as "What is the survival value of a given behavior pattern?"

In spite of the fact that ethology and neurophysiology have reached their present state over different paths it is now clear that the general questions facing them are almost identical—how and why do animals do what they do? I am convinced that this confluence of common concern, or rather, that a general recognition of its presence, is a very auspicious development for both fields. This conviction stems from personal experience, for I know that my own research direction had more meaning for me after reading the IV Symposium of the Society of Experimental Biology (1950) and first hearing about some of the ideas current in Niko Tinbergen's Oxford Laboratory at that period. This essay examines some phenomena encountered in my own work that would be meaningless if the behavior of nerve cells was not related to the behavior of the animals of which they are a part. As any biologist knows, this also means examining the differences.

III. Neuron Signals and Interactions

Neurophysiology deals mostly with the behavior of neurons (nerve cells) either singly or in small interacting groups. The magnitude of the problem it faces in relating neuron behavior to animal behavior is best appreciated in terms of an analogy. The problem may be likened to that of attempting to estimate the foreign policy or national posture of a country when the only source of information is conversations held with a few of its inhabitants, selected mostly from the general population. True, command neurons and coupled oscillators triggering coordinated contractions of assorted muscle groups to produce certain simple actions are now being described (Davis and Kennedy, 1972; also Section VI of this article), but in terms of my analogy these correspond only to administrators and military officers of the lower ranks. We do not know whether there are neurons analogous to cabinet ministers or whether solutions leading to united action of the whole animal are reached by some extraordinary consensus.

The physiological problem becomes even more awesome if the social analogy is embroidered with numbers. The human nervous system has been estimated to contain 10^{10} to 10^{11} nerve cells—several hundred times the population of the earth. The nervous systems of insects are thought to contain about 10^5 nerve cells—numerically equivalent to the population of a small city. For some of us this justifies the tendency to turn to invertebrate animals for answers.

A. THE NERVE CELL MEMBRANE

On the more positive side, neurophysiology has now attained a fair understanding of the "grammar and syntax" of communication between individual nerve cells. This knowledge is built on a solid logical foundation laid by a series of discoveries over the past half-century that culminated in the work of Hodgkin (1951; reviewed in Hodgkin, 1964) and his associates.

These discoveries concerned the biophysical properties of the membranes that serve as the living "skin" of single nerve cells. The membrane of a neuron at rest is able to extrude (or exclude) from its interior certain ions, particularly positively charged sodium ions, that are present in high concentration in the body fluids. At the same time the membrane is able to accumulate other similarly charged but less common ions, notably those of positively charged potassium. Physics predicts that ions, if free to move, will migrate "down hill," that is, from a region of high concentration to a region of lesser concentration. Therefore, this segregation of ions by the living membrane tells us that it is doing work, that is, expending energy in sustaining a physically "improbable" distribution of charged particles between its interior protoplasm and the fluid that surrounds it.

The distribution of negatively charged ions of other species as well as of water molecules is involved in this situation. Details cannot be given here and those interested in a precise account should consult Hodgkin (1964) or Katz (1966). It is sufficient to point out that the metabolic energy expended by a living, resting, nerve cell in sequestering some ion species and excluding or extruding others places its membrane in a state of electrical tension. This is because the unequally distributed ions carry electric charges. The net tension is manifest as a potential difference, the membrane potential, in which the outside of the membrane in contact with tissue fluid is positively charged with respect to the cell contents.

This state of electric tension is sustained unchanged so long as the cell is living but inactive. If the membrane of a nerve fiber is invaded by a nerve impulse these special ion-segregating properties become altered abruptly and for about one-thousandth of a second. Selectivity for sodium ions momentarily disappears and sodium ions pour inward and down their

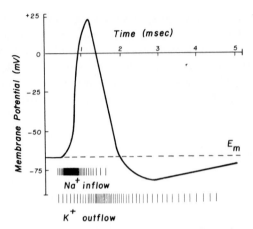

Fig. 1. Diagram showing the relation between the flux of certain ions across the axon membrane and sudden changes in membrane potential known as the action potential or spike. Vertical axis, potential of the inside of the membrane with reference to the outside; horizontal axis, time in milliseconds. Depolarization and reversal of the sign of the membrane potential is due to an influx of positively charged sodium ions (Na^+). This influx terminates and is replaced by an efflux of positively charged potassium ions (K^+) that eventually restores the membrane potential to its resting level (E_M). (After Roeder, 1967a.)

diffusion gradient, to be followed by an outward migration of potassium ions. These mass ion fluxes—positive inward followed by positive outward—cause a transient change in the membrane potential. From being positive outside, the membrane potential drops to zero and then briefly reverses its sign before it returns to a resting value (Fig. 1). This change in potential can be registered by an electrode tip placed either within the membrane or near the exterior of the nerve cell. It is known as the action potential, spike potential, or simply spike.

The so-called Hodgkin-Huxley interpretation of membrane events in a neuron is now supported by massive quantitative evidence. It has provided a foundation for modern neurophysiology much as was done for ethology by Darwinian evolution. Its accounts for the propagation of impulses along nerve fibers (axons) as well as for their genesis in nerve cells through synaptic contacts with other nerve cells.

B. Impulse Propagation

It has long been known that if the membrane potential of an axon is lowered, that is, if the outside of the membrane is made more negative through a short circuit caused by injury or by an electric current passed through it from an external source, the membrane becomes electrically unstable and liable to the self-perpetuating breakdown described above. When a critical degree of depolarization (reduction in membrane potential) is brought about the special sodium-excluding properties begin to collapse. Since the "ionic avalanche" that ensues amounts in itself to a local short circuit in the membrane, neighboring regions of membrane begin to discharge through this electrical "hole," become depolarized and "go critical" in their turn, and generate an action potential. Thus, an action potential propagates itself over the electrically excitable membrane of the nerve fiber after the fashion of a spreading forest fire or mountain avalanche.

The velocity with which a spike propagates over the membrane depends in part on how rapidly this critical loss of external positivity (depolarization) can take place and precipitate an "ionic avalanche" at each successive point on the membrane. It should be noted that depolarization, a decrease in the local membrane potential, precipitates an action potential not hyperpolarization or an increase in membrane potential. This is significant in understanding how nerve impulses are naturally generated by nerve cells *in situ*.

C. IMPULSE GENESIS AND INTEGRATION

It is hard to reach a proper perspective in relating these events to those to follow because of the enormous range of dimensions in which they take place. The ion movements described above cover distances far below the microsocopic level and in a membrane only a few hundred Ångstrom units in thickness. This membrane is the active "skin" of nerve cells whose cell bodies are measured in micrometers and thus fully visible under the microscope. From the cell bodies extend axons or nerve fibers that may be up to a meter or more in length and have diameters ranging from less than a micrometer to a few tens of micrometers. Axons are long narrow cylinders of electrically excitable membrane, and action potentials, once initiated, sweep over their full extent. Axons are the medium of fast communication in the nervous system and spikes are the message units. Spikes are generated in trains and bursts at synaptic contacts with the cell body and dendrites (fine cell processes) arising nearby. Spike patterns constitute the message. The main concern of this subsection is the manner in which such messages are generated.

The events described in the preceding subsection have mostly been studied in axons isolated from the cell bodies of their neurons. In an intact nerve cell the axon is concerned solely with the propagation of an already organized train of spikes to some distant point in the body. The receptor membrane where genesis of this train takes place has properties somewhat different from that enclosing the axon. In sensory nerve cells receptor membrane may convert external changes that are optical, mechanical, or chemical in nature into depolarizations capable of generating spikes. This mechanism is little understood.

Further "downstream" in the meshwork of the central nervous system, impulse trains in sensory neurons impinge on the dendrites or cell bodies of second-order interneurons, and these in turn interact with others. The trail is lost until it is picked up at motor neurons that generate and transmit spike patterns to muscles and glands. Motor neurons have been extensively studied from the time of Sherrington (1906) to that of Eccles (1957). Much less is known about the behavior of interneurons that feed impulse trains into the receptor membrane of motor neurons from many directions.

Sherrington called motor neurons "final common paths" because they deliver impulse sequences to groups of muscle fibers and these collectively determine the "shape of the action." The pattern of impulses generated by each motor neuron is determined by integration or adding together of the effects produced by impulses arriving in interneuron axons converging from the neuronal mesh upstream.

Membrane on the dendrites and cell body of vertebrate motor neurons receives hundreds of close extracellular contacts or synapses made by the axons of sensory neurons and upstream interneurons. In most of the invertebrate animals the synaptic area of internuncial and motor neurons is generally set off together with the root of the axon on a T-shaped side branch from the cell body.

Chemical synapses probably form the majority of neuron junctions responsible for the genesis of behavior. In these the membrane of an axon termination (the presynaptic element) lies closely apposed to but not continuous with a small area of receptor membrane (the postsynaptic element) belonging to the downstream neuron. Ionic disbalances similar to those in axons maintain a membrane potential across both pre- and postsynaptic membranes, but beyond this point the synaptic mechanism differs from the axonic mechanism in several important respects.

Depolarization of the presynaptic axon termination by an arriving spike causes the release across its membrane of small packets of a specific chemical mediator. Acetylcholine and norepinephrine are established chemical mediators and several other substances are candidates. After a millisecond or so the postsynaptic membrane of the recipient neuron responds electrically to the arriving mediator in one of two ways that are specific for the synapse in question. First, it may depolarize because the mediator causes a general increase in permeability, allowing a number of ion species to travel more readily down their diffusion gradients. Since the positive sodium ion is in greatest abundance outside, the net result is an inward movement of positive charges and a local depolarization. This means that this type of unit synaptic event is potentially excitatory, but it must be noted that in the type of receptor membrane in question an electrogenic spike cannot be supported so that the local depolarization merely spreads decrementally over adjacent membrane and may die away in a few milliseconds without generating a spike. If, however, a number of excitatory synapses undergo local depolarizations, and if these depolarizations overlap to a critical degree in time as well as space on the receptor membrane, the net depolarization produced (excitatory postsynaptic potential or EPSP) may be sufficient to cause the genesis of one or more propagated spikes at the root of the axon.

In assessing this rather complex situation it is important to emphasize that the adding together of local transient potentials at individual synapses has both temporal and spatial dimensions. Each local depolarization produced at a synapse by an arriving presynaptic spike dies away after a few milliseconds so that there must be a critical incidence of spikes *in time* if the resultant EPSP is to reach a magnitude sufficient to trigger a propagated spike in the axon; also each local depolarization becomes smaller as it spreads over adjacent membrane so that there must be a critical concentration of local depolarizations *in space* on the receptor membrane if the EPSP is to have sufficient magnitude to generate a spike. These two parameters of synaptic integration are known as temporal and spatial summation.

Chemical synapses may respond to the presynaptic release of mediator in a second way—by becoming hyperpolarized. The mediator in this case causes an increase in ion permeability only to certain ion species such as chloride and potassium. Negative chloride ions are more concentrated outside the receptor membrane and positive potassium ions more concentrated inside. Therefore, there is a flux of positive ions out of and negative ions into the cell. These migrations momentarily increase the external positivity (or internal negativity) and thus add to the already existing membrane potential. The membrane is said to have become hyperpolarized.

These local hyperpolarizations act to offset the depolarizations produced at excitatory synapses. Local hyperpolarizations produced at inhibitory synapses summate in space and time in a manner similar to depolarizations produced at excitatory synapses, differing from them principally in the *sign* of the effects produced. Since both are local nonpropagated events and limited to the receptor membrane of the nerve cell they sum together in algebraic fashion and the net membrane potential change "seen" by an electrode inserted into this portion of the nerve cell depends on which predominates. If a majority of the impinging spikes arrive at inhibitory synapses the receptor membrane hyperpolarizes, that is, it moves further from the point of critical instability. This potential change is known as the inhibitory postsynaptic potential (IPSP).

In attempting to visualize this rather complex situation (Fig. 2) it must be emphasized that the region of receptor membrane of a vertebrate motor

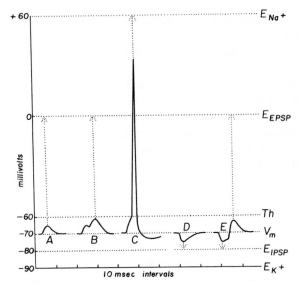

Fig. 2. Diagram of changes in the membrane potential of a postsynaptic neuron caused by the incidence of presynaptic impulses. Under resting conditions the membrane potential (V_M) is about 70 millivolts, negative inside the cell (vertical axis). A, an excitatory postsynaptic potential (EPSP) is generated by the arrival of presynaptic impulses. During the EPSP there is a general increase in ion permeability that tends to depolarize the membrane (vertical dotted line). However, the EPSP in this case is insufficient to reach the "critical" level of depolarization (horizontal dotted line, Th) needed for the genesis of a propagated spike. B, a second EPSP follows shortly on the first, arriving, however, during its falling phase. The spacing of the two EPSPs does not allow sufficient temporal summation to generate a spike. C, the second EPSP follows the first more closely and the two together summate so as to permit genesis of a spike. The rising phase of the spike is due to sodium ion influx and proceeds (as in Fig. 1) toward sodium equilibrium (horizontal dotted line, E_{Na}) but stops short owing to efflux of potassium ions. D, an inhibitory postsynaptic potential (IPSP) caused by presynaptic impulses reaching inhibitory synaptic terminals. The membrane potential tends to increase (hyperpolarize) toward another point of equilibrium (dotted line, E_{IPSP}). The IPSP then decays decrementally to the resting level (V_M). E, interaction of an IPSP with a subsequent EPSP equal to that generating a spike in C. The summation of IPSP with EPSP in this case prevents the EPSP from reaching a critical level of depolarization. (From Roeder, 1967a.)

neuron is specifically sensitive to excitatory and inhibitory mediators which affect its permeability in different ways. In this respect its receptor membrane is quite unlike the membrane that forms its axon and propagates spikes. One can visualize the receptor membrane as being exposed to a barrage of unit hyper- and depolarizations caused by spikes arriving at dozens or hundreds of synapses and spreading decrementally over its surface. If the net effect at any instant is a critical degree of depolarization this draws sufficient current from the root of the axon to depolarize it and trigger one or more propagated spikes.

Some thought about this integration taking place at the surface of a single neuron reveals that it has considerable significance in the mechansims of behavior. At the same time it is rather challenging to realize that this complex set of interactions may be repeated hundreds of thousands of times at different levels of organization in the nervous system of an intact animal. Elsewhere (Roeder, 1967a) I have likened a motor neuron to an administrator in an administrative hierarchy. He receives continuously from "above" a multitude of decisions, both positive and negative, and his responsibility is whether or not to act at a given moment. It is some comfort that action is simple and easy to interpret in the case of the motor neuron—one or more propagated spikes that travel down the axon to its termination.

Little is known about synaptic transactions in the mesh of interneurons that make up the bulk of the central nervous system. Many have highly specific and complex dendritic architectures, suggesting specialized modes of integration. Several chemicals are suspected of serving as excitatory and inhibitory synaptic mediators. In certain neuronal systems, particularly those concerned in rapid action such as evasive behavior, some synapses are electrical rather than chemical. In these the pre- and postsynaptic membranes are in even closer apposition so that current flow caused by an arriving presynaptic spike can directly influence the postsynaptic membrane potential. The adaptive significance of electrical synapses seems to be that no time is lost because of the synaptic delay occupied by diffusion of mediator across the synaptic cleft. This might be an important factor in evading a predator.

Other variations can only be listed here. Inhibition can also take place through presynaptic mechanisms. It is probable that certain interneurons can also interact with one another electrotonically, that is, without the intervention of propagating spikes. The excitable cells of some species depend on ions in the body fluids other than sodium and potassium, notably calcium and magnesium, as vehicles for the charge transfer required for membrane potentials and spike propagation.

D. Hierarchies

The student of animal behavior is justified in throwing up his hands at this point, if indeed he has reached it. He has been asked to assimilate details about the affairs of single nerve cells but is given no clue as to how all this relates to behavior. Is this constellation of neurons as incomprehensible as the star patterns of the night sky? Or is there a relationship at least as finite as that existing between the actions of individual citizens and the foreign policy of their land?

The organization of a few neuronal hierarchies has been intensively studied, but these are concerned either with sensory or with motor mechanisms and the organization of the neuronal mesh in between remains mostly unknown. The most complete sensory studies concern the vertebrate visual system (Lettvin et al., 1959; Hubel and Weisel, 1965). Individual visual receptor cells merely generate a potential change when light falls upon them. Spikes generated in nerve fibers coming from a field of such receptor cells are sorted and ordered by several levels of higher-order interneurons. Images are sharpened by lateral inhibition in which excitation of a given receptor cell causes suppression or lowered excitability of receptor cells that immediately surround it. This effect enhances contrasts at the edges of objects viewed. Interneurons can be found in the brain that respond specifically and only to whole fields, or points, or edges, or corners and lines having fixed orientations, or only to movement in one direction or another.

On the motor side (Sherrington, 1906; Eccles, 1957), motor neurons supplying the various groups of muscle fibers (motor units) that make up a muscle are gathered in the spinal cord into motor neuron pools. The number of motor units activated in a pool determines in part the tension exerted by the whole muscle. The motor neuron pools of antagonistic muscles (e.g., extensors and flexors) are coupled by inhibitory interneurons that prohibit simultaneous contraction. Motor neuron pools supplying synergist muscles are similarly coupled by excitatory interneurons. These complexes are in turn subject to control by command interneurons whose activity determines coordinated reflex (e.g., postural) and voluntary movements.

Both the sensory and motor systems include extensive regulatory or feedback neuron loops whose sense organs (proprioceptors) measure how much of the action has been accomplished and how much still remains to be performed. These ubiquitous feedback systems include synaptic interactions that are both excitatory ("turning-on") and inhibitory ("turning-off").

Before leaving this cursory survey some of the major gaps in neurophysiology are worth noting. The sensing mechanisms and organization of the neuronal hierarchies concerned in the chemical senses—olfaction and taste—are much less well understood than those subserving vision and hearing. In all the senses we have no firm conception how receptor membrane of a sense cell transforms an external change into a potential capable of regulating the performance of an interneuron or triggering a spike. The minimum external change capable of this may by a quantum of radiant energy, a displacement of less than one Ångstrom, or a single molecule of an active chemical, but the potential change that results probably involves the transfer of tens of thousands of ions across the receptor cell membrane. Therefore, this first small step in the behavioral mechanism involves not only the trans-formation of one energy form into another but also enormous amplification by a mechanism that remains completely mysterious. One can only visualize a valvelike mechanism exerting this form of control. Finally, and of most interest to the student of behavior, there is no hint of the mechanism whereby an animal selects from its behavioral repertoire a behavior mode that is adaptive under a given set of conditions.

IV. Some Reservations

Before attempting to narrow the gap between neurophysiology and ethol-ogy it seems necessary to discuss some of the reservations expressed infor-mally by ethologists about the value and relevance of investigating the mech-anisms of behavior. What can be learned about a behaving animal by taking it apart? Is there heuristic value in trying to combine the ideas and view-points of both fields?

A. Adaptedness

Physiologists are somtimes accused of having relatively little concern for the adaptedness or survival value of a given physiological mechanism. I be-lieve that many physiologists take these matters for granted, but when they are lost sight of, as they sometimes are, there seem to me to be three explanations.

First, if physiology originated as an adjunct of medicine (Section II) then its early concern was with a single species. In this context comparative ques-tions about adaptedness and survival value would have had little relevance. However, the growth of comparative physiology over the past century has greatly changed this situation (e.g., see Prosser and Brown, 1961; Bullock and Horridge, 1965). Second, ethology begins with the actions of animals in their natural environments—the arena where evolution is actually taking place. Most of the actions that concern physiology take place "below the skin," that is, at a level once removed from this arena and submerged in the regulated environment of the body. Of course, physiological mechanisms must have survival value and be subject to selection, but this becomes less obvious when only certain of their component parts are studied in isolation. The brakes of a car have clear survival value but this becomes less obvious when a brake cylinder is considered by itself. The third point is related to the second, and derives from the well-known fact that as one narrows one's perspective from comparison of whole animals to comparison of their internal organs, their tissues, their cells, and finally their biochemical processes one finds that species which are clearly distinct at the whole-animal level cannot be discriminated at the internal-organ level while phyla become indistinct at the tissue and cellular levels of organization and a great many biochemical systems are common to all living things. Thus, questions of adaptedness and survival value take another form when living things are studied at finer levels of organization.

B. Instrumentation

Ethologists accustomed to face-to-face contact with their subjects are inclined to feel that the living system may be submerged or at any rate distorted by the elaborate array of complex "gadgetry" used by physiologists to extract information from the nervous system. In part this reaction may be dismissed as inevitable when one first encounters a strange technology. However, the complex instrumentation of neurophysiology is needed in order to amplify the small and brief events in neurons and then to transduce them into a physical form (light, sound, etc.) detectable by human senses. It is indeed a curious fact that man seems to be relatively insensitive to external electrical events and yet is prodigiously sensitive to certain electromagnetic radiations, mechanical displacements, and certain chemicals. It might be thought that this is because electrical sensitivity of a mechanism whose operation is in part electrical would result in behavioral chaos, but an electrical sense has not been omitted from all animals. Electric fish are highly sensitive to minute electrical gradients in the water (Lissman, 1958; Kalmijn, 1971).

Returning to the question of the elaborate instrumentation, there is indeed some danger that it may dictate the direction of a research program rather than serving merely as a means to an end, particularly if it has cost a great deal of money! But physiologists are not alone in being prone to this dictation. Ethologists must experience it also, particularly as they become equipped with more esoteric means of observing, tracking, and recording the actions of animals.

C. Precision

The fallacy that neurophysiology is more precise than ethology in its methods is believed by some neurophysiologists as well as ethologists. The precision obtainable in measuring the performance of a system is limited by the accuracy of the measuring tool and by the degree to which the variables can be controlled. If one is content to study a banal problem it is possible to do this with great precision. On the other hand, the methods available for studying the really interesting problems, that is, those "out on the edge" of a given field, are invariably imprecise, and may reveal a trend only after many measurements have been averaged. Examples are the nature of pulsars in radioastronomy and of ultimate particles of the universe in particle physics, two fields commonly considered quantitatively precise by the layman.

Precision in detection and measurement is then a matter of the state of the art and hence is subject to improvement. Precision limited by the number of uncontrollable input variables of a system presents a far more perplexing problem and one that besets both ethologists and neurophysiologists. The major input variables controlling the behavior of an axon *seem* to be known and controllable when the system is regarded superficially (Section III), but closer inspection shows that there is not universal agreement as to their number and nature. At higher levels of neural organization—systems made up of whole neurons, two neurons, and neuron meshes—the variables multiply and the predictability of the system drops in proportion. The complexities of animal behavior are greater than these by orders of magnitude, and if it were not for the unifying principles of evolution ethology would probably still be entirely empirical. From this it follows that precision is a relative term and that it is perhaps best regarded as a habit of mind in the investigator rather than as a characteristic of the field being investigated.

D. Objectivity

It used to be thought that one could be objective in regard to a physical measurement such as the speed of light, but Einstein showed that this is not necessarily so. It is sometimes thought that neurophysiology is more objective, or rather that one is capable of being more objective in its study, than is the case with ethology.

Actual objectivity in science is, in my opinion, never achieved, at least by practicing scientists. The best one can do is to have some realization of one's own subjectivity regarding one's interest. Objectivity accompanies the naïvete with which one first observes a novel phenomenon—a neural or behavioral event or perhaps a pair of events. But at this instant the curious mind cannot avoid being invaded and committed to a hypothesis, be it about mechanism, causation, or survival value. Thenceforth one's judgments are subjective, for, in spite of conscientious efforts to seek evidence pro or con the hypothesis, there is always a subvert bias in favor of evidence promoting the survival of one's "brain-child." But this is not entirely the point: more important is the fact that the hypothesis serves to polarize one's thinking along a certain mental axis, blinding one to evidence suggesting an explanation lying on an entirely different axis. One of the most stimulating experiences in research is when a mountain of evidence finally forces one to discard a pet hypothesis and to become enthralled by a new "brain-child."

All this is self-evident to those who have spent much time in research, but the point is that it does not permit an objective view of one's own work, whether it be with nerves or whole animals. The best that one can hope for is that at a certain level of hypothesizing another investigator will reach similar conclusions from other evidence. I believe that a scientist convinced of his own objectivity has reached the end of the road. No more progress is possible.

It seems to me that subjectivity towards one's science and its subjects derives from a sense of their beauty. Exhaltation may be found in a first glimpse of some of the "order within disorder" of evolution, in the perfection of a mimic in nature, in seeing the information content of a spike train, and last but not least, in the "infallibility" of one's very own hypothesis! One can fall in love with one and all of these and become its willing captive. In rare moments of objective reflection I have asked myself why I continue to do what I do and this is the only answer I can find.

The dangers (and indeed the humbug) attendant to this attitude are many and obvious. For instance, I find it less "painful" to carry out a neurophysiological experiment that destroys a dowdy moth than one that employs a gaily colored species. But even this childishness pays a dividend—I am doubly careful that the experiment with the "beautiful" moth yields a maximum amount of useful information. Also on the debit side, a hypothesis built on human notions of mechanics and economy has actually held me back from understanding how a particular biological mechanism reached its present condition through evolutionary processes (Section VIII).

But most scientists would find their search too mundane if they lacked subjectivity and were not captivated by its attendant visions of beauty. Some might not admit this, but without it I cannot see how curiosity can flourish.

V. INSIDE THE SKIN

Elsewhere (Section II) I have suggested that the interests of ethology and neurophysiology can be said to overlap at the skin of an animal. The term "skin" must be taken figuratively and in its broadest possible context so as to include all external sensory surfaces and skin derivatives as well as all actions of an animal that can be observed from the outside. In this sense ethology can be said to deal with events outside the skin; neurophysiology with events inside it.

In this section I shall attempt to suggest one way in which neurophysiological concerns can find contact and merge with ethological concerns at this interface (Fig. 3). Of course, this is only one among a number of possible ways of asking questions about behavioral mechanisms that take into account their adaptive value in natural surroundings.

Analysis of a behavioral mechanism cannot, by definition, be carried out under natural conditions. This means that as the level of analysis becomes finer its ethological significance recedes as its physicochemical significance becomes more evident. One object of the scheme is to show how this progressive loss of contact with the ethological domain can be lessened if questions relating to the ethological situation "outside the skin" are formulated at as many levels of the analysis as possible. This provides the whole study with a form of feedback that prevents it from "running-away" from biology as it approaches the molecular levels of science. A second object of the scheme is to illustrate for ethologists some of the paths that might be taken by a physiological study of a behavior pattern. Thus, the scheme is not to be taken as a model of a nervous mechanism but merely as a suggestion of the order in which the elements of such a mechanism might be examined. The afferent side only is considered because this seems to be sufficiently illustrative.

The study begins in the ethological domain "outside the skin" with the observation of a given response of an animal to a stimulus configuration likely to be encountered in its natural environment. At this stage, using the behavioral response as criterion, parameters of the stimulus configuration

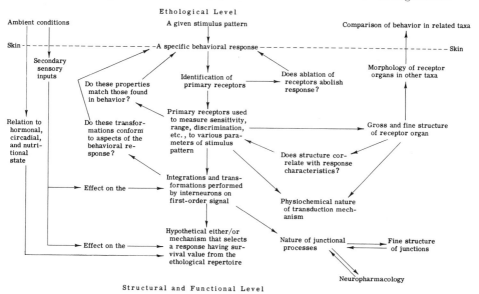

Fig. 3. A scheme suggesting relations between neurophysiological and ethological observations and experiments across the hypothetical skin of an animal. For explanation see text.

are tested separately and as nearly as possible under field conditions in an attempt to identify those having special signal value. These ethological observations suggest physiological observations "below the skin" (dotted line).

Simple elimination or ablation experiments are carried out in order to identify primary receptors or receptor fields if these were not already apparent. These lead to more incisive neurophysiological studies both of the compound electrical response or electrogram and of unit electrical responses of single receptor cells recorded both from first-order afferent fibers and directly from the receptor cells themselves. This is sensory physiology, an important branch of neurophysiology. One of its objectives is to measure the discriminatory capacity, spectral range, and sensitivity of the receptor organ. If there is a large receptor population and the behavioral experiments suggest fine discrimination this phase of the study will occupy a considerable research effort.

Ideally, the order in which these and following studies are pursued should follow the order of signal processing by the nervous system, but this is not always possible for technical reasons. Often the signals most accessible to neurophysiological methods are spikes transmitted centripetally by higher order interneurons. These interneurons commonly perform integrations and transformations on the signals from sense cells, regrouping these in various ways. These integrations may be spatial and temporal and involve inhibition as well as excitation. They are most interesting because the regrouping of afferent spikes usually accentuates certain aspects of the stimulus configuration while discarding others. A well-known example is lateral inhibition in compound visual organs whereby excitation of one receptor cell tends to suppress the responses of adjacent receptor cells. This tends to sharpen the total pattern conveyed to the central nervous system, accentuating lines and edges in the field. Similar phenomena having temporal rather than spatial dimensions occur even in quite simple acoustic detectors (see Section VII). The importance of such patterning systems suggests special synaptic arrangements between first-order and higher order interneurons of the afferent system. These arrangements must eventually be studied in detail.

This afferent pathway is postulated to converge with those subserving other sense modalities carrying information about ambient conditions or information that might conflict with or tend to suppress the given response. One must presume that the central nervous system contains what may be called an either/or mechanism that selects from the animal's behavioral repertoire a response pattern that has most survival value under given circumstances. In spite of its intense interest such a mechanism is at present merely a matter of speculation.

To the right of Fig. 3 are suggested some of the connections these functional steps might make with other areas. The most important of these aims to provide a structural basis for the operation of the sensory mechanism. The anatomical study correlates closely with the physiological findings, but it also follows a path of its own (upward arrows) that leads to comparative anatomical studies in related species and finally correlates "outside the skin" with comparative ethology. In addition, both the functional and structural studies converge downward to basic questions regarding the physicochemical properties of the receptor cell that enable it to transduce a minute amount of energy or of a chemical into an effective membrane depolarization or hyperpolarization.

Questions of the sort indicated to the left of the scheme are equally significant, and once more originate in the ethological domain. It is well known to ethologists that most types of behavioral response to a specific stimulus, that emanating from a mate, or a predator, or food, for instance, do not occur in a complete form unless environmental or biological circumstances are "right." Therefore, an understanding of these ambient circumstances is of some importance. In a physiological study it is nearly always necessary to enclose the animal in the "controlled" environment of a laboratory, thereby changing an unknown number of the environmental circumstances even if some of the biological circumstances such as nutritional state or endocrine balance are accounted for. The descending arrows to the left of the scheme admit to this complex matter and suggest that where it is possible to take such circumstances into account their action on the integrating and choice systems is worth studying.

Many more arrows could be drawn and details filled in, but I feel that they would only obscure what I consider to be the main point of this scheme. This is suggested by the arrows ascending at several levels from the central analytic path of the study to connect with the specific behavior pattern "outside the skin." The questions on these ascending arrows serve as a form of feedback or regulation that counters the human tendency to follow only the analytic (downward) direction of the investigation. "Taking-apart," such as the disassembly in childhood of one's first watch, is an intrinsically satisfying occupation, but it begins to approach science only if each step in the analysis can be causally related to prior steps and to the operation of the intact system. The ideal solution is, of course, actual reconstruction of the original system, but this is rarely possible with the complex "Humpty-Dumpty" mechanisms of living matter. A substitute is the habit of asking questions at each stage of the analysis that relate to an earlier stage or to the original behavioral observation. This often brings to light details and observations that seemed previously to be irrelevant or might have been overlooked. The scheme suggested in Fig. 3 contains several such questions that may arise at different levels of the neurophysiological study and be directed back to the ethological observations.

The preceding sections of this search for correlations between ethology and neurophysiology have dealt mostly with generalities. The following sections will discuss specific cases in which the behavior of nerve cells seems relevant to certain types of behavior of the animals containing them. It may be of interest to relate some of these cases to the scheme suggested in Fig. 3.

VI. Spontaneous Activity

Living things are open thermodynamic systems. Energy stored in sunlight or chemical bonds streams through them and is degraded to theat and waste products. The energy is entrapped in the chemical bonds of substances synthesized for growth and repair and stored likewise in substances from which it can be released in the form of electrical and mechanical work. During life, energy utilization is a continuous process in respect to growth, repair, and storage; energy release is mostly discontinuous as it is manifest in the muscle actions of animals and in the action potentials of nerve cells.

A.   The Threshold Concept

Lack of movement in a healthy animal or lack of impulses in a healthy neuron indicates that these systems are in a poised rather than a passive state. Like a loaded gun, each needs only to be triggered to release some of its potential energy. In both the magnitude of the energy release, that is, the amount of activity of the animal or of current flow across the axon membrane, bears no necessary proportional relation to the magnitude of the triggering stimulus; this is a function of the internal state of the living system.

This poised state justifies the well-known sensitivity measure of "threshold." A system requiring a large trigger stimulus is defined as having a high threshold; a more irritable system has a lower threshold. This means that excitability is measured in terms of the reciprocal of the trigger stimulus required. The threshold concept is applicable in those examples of animal and nerve cell behavior where the poised state is relatively stable and the response is all-or-none.

For example, an axon lying in its natural medium but separated from its cell body will not propagate an action potential over its length until a trigger stimulus locally depresses its membrane potential to a critical level (Section III,1). The magnitude of the stimulus needed to cause this critical depression measures its threshold. If external factors are held constant the threshold is determined by the local state of the axon membrane and bears no relation to the parameters of the action potential subsequently propagated from the point of stimulation.

A predator that attacks its prey by striking from ambush may also be said to have a threshold. The probability that it will strike within a specified time interval is determined by the suitability and accessibility of the prey and by the predator's internal state. The vigor of the strike is independent of these factors. For instance, a partially surfeited predator or one presented with suboptimal prey would not be expected to make a partial strike that falls short of its target. The strike threshold of the predator depends upon its internal state and on prey characteristics.

B. The Role of Spontaneity

These situations of the isolated inactive axon and the poised predator are in fact special cases when their total behavioral repertoires are considered. The initial immobility (inactivity) and the all-or-none nature of this type of behavior has made it particularly amenable to experimental studies, and these have naturally emphasized the threshold concept. In describing and measuring other modes of animal and neuron behavior this concept is less useful.

For example, when a predator is deprived of food it will eventually hunt rather than lie in ambush, or if it is never given the opportunity to strike at live prey even when satiated with food it may exhibit phantom or "play" strikes, as in the well-known case of the pet starling described by Lorenz (1937). If an axon remains connected to its natural spike-generating cell body it may be found to transmit a series of spikes originating in the dendritic region or cell body in the absence of detectable presynaptic or sensory input. This happens more frequently if the neuron is still part of its neuronal mesh. Even completely isolated axons can be made to generate a succession of spikes if placed in a medium deficient in calcium ions. Such activity is termed spontaneous, endogenous, or free-running. Since it is not elicited by a measureable stimulus a threshold cannot be determined.

It is perhaps more accurate to say that a spontaneously active system lacks a fixed threshold. Its sensitivity sweeps from one extreme to the other. Immediately following an impulse or a completed strike both the spontaneously active axon and the predator are refractory to previously effective stimuli—their thresholds are out of reach. Then there follows a period during which their thesholds gradually decline, action being triggered by progressively less optimal or intense stimuli. Finally, their thresholds become zero when a spontaneous action takes place. The cycle of changes is then repeated. Both are examples of systems incapable of reaching a static equilibrium with their surroundings, which, after all, is characteristic of living matter.

It is quite easy to find physical analogs of spontaneous behavior in the regenerative activity of various types of oscillator. Clockwork, or better still the operation of the ubiquitous neon flashers, are familiar examples. Many years ago I built an electromechanical "gadget" to demonstrate in simple form some principles of phototaxis. Two directional photocells facing forward each controlled an electric motor driving a wheel on either side of a three-wheeled platform. In its normal positively phototactic mode each wheel of the device was connected to the opposite photocell so that it turned more rapidly when more illumination fell on the opposite side. The result was that when presented with a single light source hanging just above it the device would approach and circle back and forth below. Thinking to make the gadget more interesting, I provided it with circuitry such that a "kick in the tail" would cause the photocells to disconnect from the contralateral drive motors and remain connected to the ipsilateral motors for a period of about 30 seconds. A "kick" thus caused the device to cease circling under the light and retreat to a dark corner where it became motionless through lack of sufficient light.

Among other things, the intent of the device was also to illustrate the fact that living systems cannot reach a static equilibrium with their surroundings. However, since illumination of the photocells was required to make the wheels turn, this inactivity in darkness presented an equilibrium situation from which the device was unable to escape. The only solution was to provide the photocell amplifiers when in darkness with increasing instability until a point was reached where current was intermittently supplied to the drive motors. This caused the device to maneuver randomly (it was also capable of detecting and avoiding on contact obstacles such as walls) until chance brought it once more under the control of the light source. The addition of this analog of spontaneous activity completed the loop that enabled the gadget to always "be doing something" and to cope with its limited environment.

A flow diagram showing each of the four behavior modes (boxes) of this gadget is illustrated in Fig. 4. Appropriate stimuli (labels on arrows) would cause it to assume one of the other three behavior modes. Provided the light source was suspended above it the device was incapable of becoming permanently motionless.

C. Spontaneous Activity in Neurons

Adrian (1930, 1931) was the first to demonstrate that trains of spikes continue to be generated in the central nervous systems of insects after the nerve cords have been completely isolated. Spontaneous activity also occurs

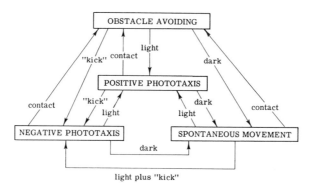

FIG. 4. Modes of behavior (boxes) of an electromechanical model. The device was designed and its environment defined in such a way that the model could never reach equilibrium (become permanently motionless). When the device is in a given behavior mode any one of three appropriate stimuli (arrows) will cause it to assume a corresponding behavior mode. For other conditions of behavior see text.

in many insect sensory nerve fibers when the receptor surface is shielded from stimuli to which the sense organ is normally sensitive. Elsewhere (Roeder, 1955, 1967a) I have discussed examples of spontaneous activity in a variety of excitable tissues and considered its genesis and adaptive significance.

At this point the ethologist may well question whether spontaneous activity in bits of isolated nerve tissue has significance other than being merely analogous to certain forms of animal behavior. Can it be demonstrated that spontanteous nerve activity is actually the mechanism behind certain behavior patterns?

Attempts by von Holst (1935) and others to answer this question immediately encountered a second question. Can an animal be said to behave after having been deprived of all sensory input? Such sensory deprivation is prerequisite to seeking an answer to the first question. Uncertainty on this point was undoubtedly responsible for some early differences of opinion and interpretation. It now seems likely that some of the confusion stemmed from the choice of complex behavior patterns such as walking in higher animals in which the surgical operation of total deafferentiation caused a catastrophic insult. In the following pages I shall review experiments that suggest positive answers to the first question. They concern relatively simple behavior patterns in insects—sexual behavior and flight.

D. INSECT SEXUAL BEHAVIOR

It has long been known (M'Cracken, 1907) that in many insects decapitation does not hinder and may even initiate oviposition. Motor patterns concerned in posture, locomotion, flight, and mating may also persist under the same conditions. The male praying mantis is sometimes attacked and eaten by the female during his approach. The loss of his head releases a complete and complex sequence of actions that enable him to clasp and couple with the female while she devours him. Should the male not be attacked during his approach these actions are not manifest until he has actually mounted the female. If a male is experimentally decapitated these copulatory movements appear with a few minutes and continue unabated for several days irrespective of the presence of a female. Thus, motor patterns associated with mating appear to be normally suppressed by centers in the male brain and are released only by stimuli emanating from contact with the female or by decapitation. In view of the female's cannibalistic tendencies this arrangement has obvious adaptive value for the species (Roeder, 1935).

The continuous nature of these rhythmic sexual movements released by decapitation of the male mantis suggested that they might be generated spontaneously, that is, without benefit of sensory input. This possibility was examined in the following experiments.

The last abdominal gangloin of the ventral nerve cord contains motor neurons controlling activity of the phallomeres, a complex set of abdominal appendages whereby the male couples with the female genitalia. The phallomeres are normally inactive in an intact male except during copulation and formation of the spermatophore. Continuous rhythmic movements of the phallomeres begin a few minutes after transection of the abdominal nerve cord above the last abdominal ganglion and persist until deterioration of the preparation.

The ventral nerve cord of a restrained male was exposed and nerves containing sensory and motor fibers connecting the phallomeres with the last abdominal ganglion were lifted on electrodes. This enabled their traffic of nerve spikes to be monitored and recorded. The nerve cord was then severed, disconnecting the last abdominal ganglion from the brain. Within a few minutes there was a great increase in the number of nerve fibers transmitting impulses, and these impulses became organized into bursts recurring every few seconds. The bursts correlated with rhythmic movements of the phallomeres and continued as long as the preparation was maintained.

This experiment leaves open the possibility that the rhythmic bursts could have been reflexly initiated and sustained by afferent impulses coming from receptors in the vicinity of the phallomeres. In a second experiment this was answered by severing all the segmental nerves connecting the last abdominal ganglion with peripheral organs such as muscles and sense organs, leaving the ventral nerve cord connecting the ganglion with the rest of the central nervous system and brain as the only possible source of neural connections with peripheral sense organs. As before, an electrode was placed on the central stump of one of the motor nerves that originally supplied the phallomeres. In this case, of course, all of the spike activity registered was efferent, that is, it could only have originated in motor neurons within the ganglion and been destined for the now separated phallomere musculature.

The level of this efferent traffic of nerve spikes was recorded first with the connection intact between the last abdominal ganglion and the rest of the central nervous system. Then the nerve cord was severed just above the last abdominal ganglion, finally and completely isolating it from any possible neuronal connection with the periphery. The results (Fig. 5) were the same as in the preceding experiment—a delayed but considerable and sustained increase in motor spikes and a grouping of these spikes into a

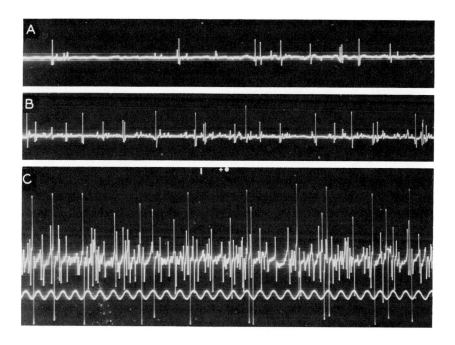

Fig. 5. Pattern of motor spikes destined for the phallomere muscles of a male preying mantis (*Mantis religiosa* L.). The phallic nerve had been severed distal to the recording electrode and all other nerves to the last abdominal ganglion had been cut. A, Before separating the ganglion from the rest of the central nervous system; B, 3 minutes after separation and complete isolation of the ganglion, a slight increase in motor spikes; G, 7 minutes after separation, many previously inactive motor units discharge spikes spontaneously during a burst. (From Roeder, 1967a.)

series of regular bursts. These motor bursts would have caused rhythmic movements of the phallomeres had they been able to reach the appropriate muscles.

This experiment provides strong evidence that the basic pattern of motor activity enabling the male mantis to copulate is initiated and organized endogenously within the last abdominal ganglion. The activity of this endogenous "oscillator" is probably modulated by impulses coming from sense organs and, as has been shown, may be completely suppressed by signals originating in the brain. Yet it is quite capable of "free-running" when released from this suppression and when disconnected from all possible sources of reflex control (Roeder *et al.*, 1960). There is evidence that neurosecretions from the corpus cardiacum play a part in the inhibitory control mechanism of this spontaneous system (Milburn *et al.*, 1960).

E. INSECT FLIGHT

Another particularly clear example of spontaneous control is found in the neuronal mechanism of wing-flapping in insects.

Insects were the first creatures capable of controlled and powered flight. Today their flight mechanisms show a number of striking muscular, metabolic, and neural adaptations (Chadwick, 1953; Pringle, 1957; Roeder, 1971). The mechanism for flapping the wings is relatively simple compared to that responsible for walking. One set of elevator muscles raises the wings; a set of depressors lowers them. These muscles constitute the basic "flight motor," which does work on the air when the elevators and depressors contract out of step with one another. The mechanism for changing the angle of attack of the wings between upstroke and downstroke to as to gain aerodynamic lift, steerage, and produce other maneuvers is a great deal more complex and will not concern us here.

In generalized insect fliers such as grasshoppers, butterflies, and moths this "two-cylinder two-stroke engine" drives both right and left pairs of wings as a single unit. A burst of nerve impulses discharged along motor nerves supplying the elevators alternates with a burst of impulses to the depressors. This cycle is repeated from 5 to about 60 times a second, depending on the wing-beat frequency characteristic of the species. The neurons producing the elevator and depressor bursts reside in the meso- and metathoracic ganglia. During steady flight, elevator and depressor bursts repeat in a fairly constant antiphase relation. Before taking flight there may be inphase contraction of the antagonists while a moth is warming up its muscles or "shivering" preparatory to becoming airborne (Kammer, 1968).

The stimuli that turn these neuronal mechanisms on and off have not been fully described. In most insects lack of contact of the feet with the ground will initiate wing-flapping, although flight may begin while this contact is still present. It may also be initiated and sustained by a head wind, movements in the visual field, certain odors, and so forth. Nevertheless, a headless and abdomenless insect can still be induced to make quasinormal flapping movements. Contact of the feet will usually terminate flapping.

It was once thought that wing-flapping is a reflex activity, each stroke being triggered and timed by signals from proprioceptors on the wings and their articulations. These signals were considered to provide phasic feedback triggering the spike bursts. Nerve centers in the thoracic ganglia are indeed exposed to a multitude of signals from such receptors as well as from the special sense organs on the head. However, the brilliant experiments of the late D. M. Wilson (1961; reviewed in Wilson, 1967) demonstrated that wing-flapping in the desert locust is driven and timed by nerve impulse patterns arising endogenously in the thoracic ganglia.

Wilson mounted locusts on supports glued to their backs and with their feet out of contact. Under these conditions the insects flapped their wings as in free flight for many hours. Using electrodes placed on nerves so as to register sensory impulses, he systematically located the various sense organs postulated by the reflex concept to provide feedback to the neuronal centers. These he eliminated one by one, also amputating the wings, legs, and abdomen and removing the viscera. Finally, all relevant muscles were severed so as to eliminate movements, and the head was amputated. This left the thoracic ganglia isolated from all incoming nerve signals.

Understandably, these deafferented ganglion preparation were somewhat unreactive. However, a brief burst of electric shocks would often initiate a sequence of levator-depressor motor spikes phased as during flight that might continue without further stimulation for some hundreds of cycles. The cycling frequency of 5 to 12 per second was less than the intact wing-beat frequency, but there seems little doubt that the neurons generating this rhythm are those responsible for generating the wing-flapping sequence of normal flight.

Wilson concluded that the timing and coordination of the wing-flapping rhythm derive from the endogenous activity of neuronal "oscillators" within the thoracic ganglia. The elevator oscillator and the depressor oscillator are coupled but may alter their phase relation. Other oscillators similarly coupled to these and also capable of phase-shift drive the muscles responsible for wing angle and steering. The excitability of this complex of neuronal oscillators, that is, the ease with which it is caused to oscillate, is modulated by the nonspecific or phase-unrelated input from proprioceptors on the wings and vibrating thorax; this input may augment the endogenous rhythm so as to produce the wing-beat frequencies attained in the intact insect.

F. Spontaneous Activity of Receptor Cells

These studies of the mechanisms of coupling and flight behavior in insects demonstrate that the basic action patterns concerned are endogenous, being composed within the central nervous system. Spontaneous activity underlies not only mechanisms generating such action patterns for it is well known that many insect sensory receptor cells are also spontaneously active. This means that when completely shielded from the stimuli to which they are normally sensitive (e.g., light, sound, chemicals) a sequence of spikes may be generated in their receptor or cell-body regions and transmitted down their axons to the central nervous system.

At first sight this spontaneous activity seems to defeat the function of a sensory receptor—to report on the presence and intensity of a specific stimulus. However, it may serve two functions; first, to heighten the general excitatory state of the central nervous system by contributing a nonspecific signal in the manner postulated by Cohen (1965), and second, to actually increase sensitivity to the specific stimulus. Increased specific sensitivity may be brought about in the following manner.

The classical code in which individual receptor cells transmit information about specific intensity differences to the central nervous system is contained in the frequency of nerve spikes traveling down their axons. Commonly, a high frequency, that is, short interspike intervals, signifies an intense stimulus. The axon of one of the acoustic receptor cells in the tympanic organ of a noctuid moth can sustain short trains of spikes at frequencies approaching 1000 per second. This frequency, equivalent to interspike intervals of 1 millisecond, registers the most intense ultrasonic stimulus the tympanic organ is capable of reporting. At the other extreme of zero stimulation the sense cell commonly generates an erratic sequence of about 5 spikes per second. One must assume that nerve spikes arriving at this low frequency, that is, when separated by intervals of 200 milliseconds or more, fail to be temporally summated by certain higher order interneurons in the central nervous systems and are therefore read as indicating "no stimulus."

A receptor neuron that discharges spontaneously in this manner actually possesses greater sensitivity to stimuli of very low intensity than does an excitable cell having a fixed threshold. This is because a stimulus of any intensity, however small, will cause a slight change in its firing frequency during the period that the stimulus is being applied. This slight frequency change against a background of spontaneous spikes may not be evident if only a single response is registered. However, if a number of consecutive responses are integrated over a given time interval by higher order interneurons a correspondingly smaller stimulus increment could be discriminated. This integration over time would not be possible if the receptor had a fixed threshold and was "silent" when exposed to stimuli ranging in strength from very weak to zero (Roeder, 1966a). The ethological implications of this type of sensory performance make it worthwhile to examine more closely the behavior of the tympanic receptor mechanism of noctuid moths and one type of evasive action that it mediates.

VII. Hearing and Steering in Noctuid Moths

The tympanic organs characteristic of this family of moths appear to be adapted specifically to detect the echolocating cries made by insectivorous bats. Each organ contains two acoustic receptor cells, one of which (A1) is sensitive to ultrasonic pulses of low intensity and will concern us here. Trains of spikes delivered by the A1 sense cells in the right and left tympanic organs provide the central nervous system of a flying moth with sufficient information to select and to hold a flight path away from the area where a bat is cruising. A full account of this and of other evasive tactics as well as of the neurophysiology of some of the mechanisms responsible is to be found elsewhere (Roeder, 1967a). The sonar system used by bats in finding their prey and avoiding obstacles in darkness has been fully described by Griffin (1958). Certain aspects of the mechanisms of acoustic evasion by moths that seem relevant to the subject of this essay will be presented and discussed in the following pages.

A. Spike Response of the Acoustic Receptor

The response characteristics of the A1 acoustic receptor of noctuids were mentioned briefly at the end of the preceding section. A typical spike response recorded from the afferent axon of A1 is shown in Fig. 6A. The stimulus (lower trace) was an artificial pulse of ultrasound having an intensity roughly equivalent to that reaching a moth from the cry made by a bat cruising at a distance of 70–100 feet. This is well beyond the range where a bat's system of echolocation would enable it to detect a moth and track its course. The moth would not, then, be exposed to immediate attack, and field and laboratory evidence (Roeder, 1966a) shows that A1 signals such as the train of spikes recorded in Fig. 6 would cause the moth to turn in flight and steer a path away from the sound source, a logical maneuver in these circumstances. It should be noted that the first three spikes in the train are separated by intervals of about 2 milliseconds, a temporal pattern that would be capable of firing the pulse-marker interneuron (see below). The remaining spikes are separated by longer and more variable intervals.

This is the simplest possible unit of information concerned in turning-away

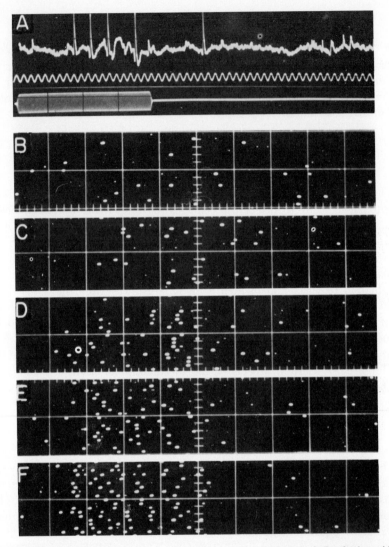

Fig. 6. A, Spikes generated by the A1 acoustic receptor of *Prodenia eridania* in response to an ultrasonic pulse (lower trace) and recorded from the neuropile in the mesothoracic ganglion. The ultrasonic pulse was 19 milliseconds in duration, of 30 kilohertz frequency, and of an intensity (+8 dB) judged roughly just sufficient to cause an intact moth to turn away from the source in free flight. B–F, The same preparation, spike sequences are reduced to rows of dots (see explanation in text), each frame containing about 20 responses at 1 per second registered serially. B, 0 dB (decibel), the lowest sound intensity used. The dots appear randomly distributed indicating that the faint sound pulses were not influencing the spike sequence. As the sound intensity is increased in small steps (by +2 dB in C, +4 dB in D, +6 dB in E, +8 dB in F) the spikes (dots) gradually "gather" into distinct groups and with shorter interspike intervals. The conventional spike pattern of frame A is one of the dot responses in F. Sine wave in A, 1 kilohertz; B–F, major vertical lines mark 5-millisecond intervals.

behavior—a single A1 spike train in response to a single ultrasonic pulse reaching one ear. However, it has been taken out of its temporal context in being "frozen" on film in order to study it. In the natural situation a cruising bat emits cries about ten times a second so that trains of A1 spikes such as this follow one another into the moth's central nervous system from both ears with this repetition rate. Therefore, the repetition rate of trains as well as the interspike intervals in each train must be taken into consideration. The patterns of dots in the lower frames of Fig. 6 are intended to take into account both of these temporal factors.

The horizontal time scale of the successive rows of dots is the same as that of the conventional record of spikes shown in A, the horizontal time span of each frame being 50 milliseconds. Each spike has been converted into a dot so that each horizontal row of dots represents a single train of spikes. One of the rows of dots in the "+8 dB" frame is identical with the train of spikes shown in A.

Stimulus pulses as in A were delivered once a second, each starting from the left-hand margin of the frame. Between each stimulus the oscilloscope beam was deflected vertically by a small amount so that each successive response (row of dots) was registered just below that preceding. Thus, each frame contains tbout 20 A1 responses, "printed" in succession rather like the lines in a verse of poetry. The difference between the frames lies only in the intensity of the sound pulses that were used. In the "0 dB" frame (B) the lowest intensity was used; in the "+8 dB" frame (F) the highest. Some measure of the intensity differences is gained by recalling that a 1 dB (decibel) difference in sound intensity can just be discriminated as such by the human ear. The "0 dB" intensity level chosen in this case was such that it failed to produce regular trains of spikes in the A1 receptor cell.

These dot patterns give a more complete though still distorted picture of how the moth ear encodes acoustic information. Both of the temporal dimensions of the response—interspike intervals (horizontal distance between dots) and interresponse intervals (vertical distance between lines of dots)—have been converted to spatial dimensions. In the "0 dB" frame (B) the dots appear to be randomly distributed during the 20 stimulus presentations. This means that their genesis bears no relation to the timing of the sound pulses, suggesting that the spikes are free-running or spontaneous, or triggered by some other stimulus. However, when the 20 responses of the "+2 dB" frame (C) are viewed as a group and compared with the frames below there appears to be a slight tendency for spike activity to "gather" at the center of the frame. This tendency would be less noticeable if fewer responses are scrutinized. The gathering of dots becomes more distinct in the "+4 dB" and "+6 dB" frames (D and E) there being not only a concentration of dots to the left shortly after the onset of the stimulus pulse but also a thinning of dots to the right, indicating some degree of adaptation, that is, a lowered tendency to fire immediately following the termination of the stimulus pulse. This is still more obvious in the "+8 dB" frame, from which the train shown in A was selected.

Perusal of the dot patterns suggests several aspects of the tympanic response that are not apparent in a single train of spikes. (1) There is no evidence of a threshold intensity for the Al receptor. As the sound intensity is increased by 2-dB (or smaller) increments there is merely a gradual gathering of spontaneously recurring spikes into trains in which the interspike intervals become progressively shorter. (2) Detection of this gathering depends on how many sequential responses are averaged, as is done of course, when one visually scans one of the frames of dots from 20 responses. (3) Though consistent turning-away behavior does not in fact take place until signals such as those in the "+8 DB" frame are transmitted to the moth's central nervous system, the patterns in the "+4 dB" and "+6 dB" frames suggest that the tympanic organ has more sensitivity than is actually realized behaviorally. One may speculate that should this "hidden" sensitivity become critical in species survival the central nervous mechanisms of the moth might become modified to integrate or average a larger number of A1 responses and in that way make it behaviorally significant. The cost of this might, however, be a longer response time. (4) Individual spikes generated by the A1 receptor are not in themselves significant information; it is the length of the time intervals separating the spikes that determines the behavioral response. This conclusion is borne out by the behavior of the pulse-marker interneuron (see below).

B. A Branching Trail from A1

The analysis of this very simple sensory mechanism (Roeder and Treat, 1957; Roeder, 1964) corresponds to the second level "inside the skin" suggested in the general scheme presented in Section V. It has led to several lines of investigation included in the scheme, two of which carried the trial back to the behavioral level. The criteria of sensitivity described above made it possible to set up a tympanic nerve preparation in the field and determine how far away it could detect the cries made by passing bats. Its maximum range was found to be 100 to 130 feet (Roeder, 1966a). A second behavioral study determined just what a moth does when it turns away in flight from a source of ultrasonic pulses. Moths were brought into the laboratory, mounted in fixed flight, and their actions examined by photographic and electronic methods as they attempted to turn away from a small loudspeaker. This revealed that a moth makes the correct decision (right or left turn) in the space of about 40 milliseconds and that it is able to do this on receiving only a single pulse, using both tympanic organs. However, a sustained turn requires a sequence of pulses (Roeder, 1967b).

Several details of the acoustic receptor mechanism are being closely scrutinized. Eggers (1928) first described its microscopic anatomy. Recently Ghiradella (1971) has thrown much light on its fine structure, using electron microscopy. In a parallel study Adams (1971) is examining the biophysics of transduction by noctuid acoustic receptors. These studies correspond to the general questions raised on the right side of the scheme presented in Fig. 3.

The "main line" of analysis of the neural mechanism concerned in this avoidance behavior, that corresponding to the central path of Fig. 3, leads to interneurons in the central nervous system whose activities are influenced by the A1 spike train and appear to bear some relation to turning-away behavior. Here the trail branches and becoms complex and its mapping is far from complete. One branch leads to the brain of the moth. I have recently (Roeder, 1970) speculated on the significance of these cerebral connections and will not consider them here. Another branch of the trail must inevitably lead to motor neurons controlling muscles that change the angle of the flapping wings. Here also the trail is incomplete. However, one type of acoustic interneuron responds to acoustic stimulation in a manner which suggests that it is directly involved in turning-away behavior.

C. THE PULSE-MARKER

A moth was restrained and prepared so as to expose the ganglia that receive sensory fibers from the tympanic organs and send out motor fibers to the muscles that steer flight. The tip of a microelectrode was passed slowly down through the mesh of nerve fibers that compose the neuropile—the ganglionic region where synaptic interactions take place. Spike patterns were sought that bore some correlation to the sequence of ultrasonic pulses directed at one tympanic organ.

Various types of correlated spike pattern were encountered, some very infrequently and others sufficiently often to make possible a closer study of their characteristics (Roeder, 1966b). As might be expected, the commonest type was identical with the spike pattern recorded in the tympanic nerve and clearly originated in the central termination of the A1 axon. The spikes shown in Fig. 6 were picked up in this region. Less commonly the position of the electrode tip was such as to register not only A1 spikes but also a spike with a longer latency and a very different mode of response to the ultrasonic pulse. The usual mode of this interneuron is a single spike in response to each sound pulse, irrespective of its duration. This earned it the title of "pulse-marker."

In the frames of Fig. 7 the small downward spikes belong to the A1 fiber of the tympanic receptor. It can be seen that they are similarly timed and grouped to those of Fig. 6A and that they continue in a train that corresponds in duration to the ultrasonic stimulus (lower trace). The larger upward spikes belong to the pulse-marker. Frames A to D of Fig. 7 were selected from a sequence of responses to a pulse of ultrasound of relatively long duration compared to a bat cry. It had an intensity judged to be only just sufficient to cause turning-away behavior in a free-flying moth.

Frames A and B show the usual behavior of the pulse-marker. The impinging A1 impulses cause it to fire once for each sound pulse and this spike discharge occurs after the arrival of the first two or three A1 impulses. In frame C the train of A1 spikes is again evident but the pulse-marker fails to discharge. In frame D two pulse-marker spikes are triggered by the A1 impulse train. A closer examination of the spacing of A1 spikes in each train indicates the synaptic mechanism responsible for these seemingly erratic pulse-marker responses.

In frames A and B the first three A1 spikes are fairly closely spaced, being separated by intervals of about 2 milliseconds; spikes in the rest of the train follow at variable but longer intervals. This suggests that the pulse-marker is brought to discharge by temporal summation of A1 impulses, in other words, that their postsynaptic effects add when the intervals between them are no greater than about 2 milliseconds. This is equivalent to the temporal summation of EPSPs illustrated in Fig. 2, and is confirmed by frame C where the first three A1 spikes are separated by slightly longer intervals and the pulse-marker fails to discharge a spike.

But the pulse-marker interneuron shows another interesting property. It fires only if the closely spaced A1 impulses are preceded by a short time interval during which no A1 spikes have occurred. Other experiments (not illustrated) used slightly stronger sound pulses that elicited a closely spaced train of A1 spikes. In this case only one pulse-marker spike was triggered at the beginning of the train. This happened even if the train continued for as long as a second. This behavior is borne out in frame D. Here the first two A1 spikes are sufficiently closely spaced to trigger the pulse-marker. Immediately after the pulse-marker spike a third A1 spike is evident, then there is a gap of about 4 milliseconds before the fourth, and then 3 milliseconds to the fifth A1 spike. However, the sixth A1 spike follows the fifth by about 2 milliseconds and this is sufficient to trigger a second pulse-marker spike. The longer intervals between A1 impulses three and five appear to have reset the pulse-marker so that it can once again respond to A1 impulses five and six since they summate over a short interval.

These characteristics indicate that the pulse-marker "takes no notice" of single or widely spaced A1 impulses such as are generated spontaneously

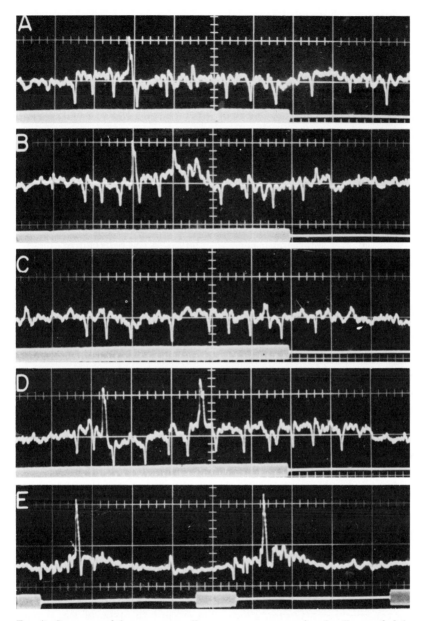

Fig. 7. Sensory and interneuron spike responses to acoustic stimuli recorded from the mesothoracic neuropile of the noctuid moth, *Caenurgina erechthea*. Smaller downgoing spikes belong to tympanic fiber from the A1 acoustic receptor; larger upgoing spikes belong to the pulse-marker interneuron. Stimulus in all frames, 30-kilohertz pulse of an intensity roughly equal to that used in Fig. 6A. In A–D the duration of the pulse was about 34 milliseconds repeated 1 per second; in E, 5 milliseconds in duration repeated 40 per second. Major vertical lines mark 5-millisecond intervals. Differences in the pulse-marker response are interpreted in the text.

by the acoustic receptor neuron. It responds only when two or three A1 impulses arrive closely spaced. At the same time, the pulse-marker takes no notice of a long train of closely spaced A1 impulses, such as would be generated by a continuous tone of sound, only responding once at the beginning of the train. However, it is capable of responding to each of a series of short pulses. In frame E sound pulses only 5 milliseconds in duration follow one another at 25-millisecond intervals, that is, with a frequency of forty per second. The short A1 train in response to each sound pulse is sufficient to trigger a pulse-marker spike since each A1 train was preceded by a period of A1 "silence."

Students of animal behavior may once again question the ethological significance of these neurophysiological nuances. Does the performance of the pulse-marker interneuron bear an relation to the behavior of free-flying moths when approached by hunting bats? The evidence for such a connection is still circumstantial, but I believe that it is quite suggestive.

The pulse-marker appears to discard or fail to react to certain aspects of the information contained in the pattern of impulses generated by the A1 receptor in the moth's ear. This property provides this simple auditory system with a means of discriminating some irrelevant sounds from those signals which it appears to have become adapted to detect. As we have seen, single A1 impulses occur spontaneously and this increases the typanic organ's potential sensitivity. Single A1 spikes may also be generated by the movements of the moth's wings during flight and by random clicks in the environment. The pulse-marker does not "attend to" this spontaneous activity or to such very brief sounds unless two or three A1 spikes occur in rapid succession. At the other extreme, the interneuronal effect of long ultrasonic pulses or continuous tones is not commensurate with their duration, in spite of the fact that such signals cause correspondingly long trains of A1 impulses. Sound pulses a few milliseconds in duration are individually just as effective as long pulses in eliciting a pulse-marker response. The "resetting" property of the pulse-marker means that short sounds repeated many times in a second are much more effective in generating pulse-marker spikes than a continuous tone lasting for the same period (Fig. 7D and E). Bat cries consist of ultrasonic pulses 5 to 15 milliseconds in duration that are repeated from 10 times a second by a cruising bat to over 100 times a second when it is tracking prey (Griffin, 1958). This is just the type of signal pattern to elicit a maximal number of impulses from the pulse-marker. Experiments with moths flying free in the field (Roeder, 1964) and in stationary flight in the laboratory (Roeder, 1967b) showed that when the insects are exposed to long tones and to various patterns of sound pulses a pattern approximating to that found in bat cries was most effective in causing turning-away behavior. It is significant that when the field work was carried out the pulse-marker interneuron had not been discovered and there was no evidence of a mechanism for this discrimination in the performance of the tympanic organ. The existence of an interneuronal mechanism such as that described above was not anticipated.

Much remains to be done. Events further "downstream" in the neural mesh of the neuropile are quite unclear although the muscle actions involved in turning-away are being studied and the neuronal mechanisms concerned in "turning-on" the evasive behavior are being worked out. These matters have been reviewed Roeder, 1970, 1971) and will not be considered here.

VIII. The Palp and the Pilifer

The latest among my efforts to link insect behavior to its neural mechanisms is more incomplete than the others summarized above. But it seems appropriate here since it illustrates some of the matter discussed in Section IV: how a heuristic hypothesis is prerequisite to an experimental study, yet how the subjectivity it confers on the experimenter can actually hinder his search for the manner in which the mechanism has in fact become organized in the course of evolution; how finally the real solution reached by the organism is so much more interesting than the man-made hypothesis. In such matters ethology and neurophysiology are on common ground.

A. Evasive Behavior in Hawkmoths

The sphingids are a small family of agile moths ranging from about the size of a bee to that of a small bat. Most species are crepuscular or nocturnal. Hawkmoths lack the thoracic tympanic organs typical of the noctuid family, and were generally considered to be deaf. An ear warning them of the approach of bats was though to lack adaptive value since most hawkmoths are fast fliers and some species are too large to be captured by the species of insectivorous bats found in the northern hemisphere. However, an early eighteenth-century observer describes the behavior of unidentified hawkmoths as well as noctuid moths (Phalaenae)) "The Sphinxes again, and Phalaenae, during the night fly about the flowers of the marragon [sic, = tarragon?] and other lily plants emitting an agreeable smell; during the night, scarcely a voice could be raised than they would turn around very swiftly, and the antennae appear, as it were, convulsed" (Bonsdorff, 1709).

In the spring of 1967, shortly after informing an audience in Austin, Texas, that hawkmoths were unable to hear, I stood before a hedge of jasmine at dusk. A dozen or more hawkmoths of the species *Celerio lineata* F. (the white-lined sphinx) hovered in front of the blossoms with outstretched tongues (Fig. 8). From habit rather than from conviction I jingled my bunch of keys—a rough and ready source of ultrasound. The moths, which are very rapid fliers, seemed to disappear abruptly in the half-light, only to return and resume feeding in a minute or so. Half-a-dozen specimens were captured and brought back to my laboratory in the hope of locating the acoustic receptor.

Electrodes were used to probe various regions of the moths' nervous systems and it was soon found that exposure to ultrasonic pulses caused large nerve spikes to descend the cervical connectives. Having localized the acoustic receptors to the head, I proceeded to remove or block with wax assorted appendages and cranial sense organs, not omitting the antennae mentioned by Bonsdorff. Within a few days it was established that the labial palps, in moths large segmented appendages extending forward on either side of the coiled tongue (Fig. 9A), play a major role in acoustic reception. Acoustic sensitivity was measured, using the interneuron response in the cervical connectives as criterion. The response was found to be maximal to the sound frequencies present in the cries of bats. Sensitivity dropped 100-fold when both labial palps had been amputated or deflected laterally on their articulations with the base of the cranium.

These preliminary findings suggested a hypothesis regarding the mechanism for detecting sounds (Roeder et al., 1968; Roeder and Treat, 1970). Only after 3 years devoted to numerous attempts to test this hypothesis did a chance observation show that the acoustic mechanism operates on a different and unsuspected principle (Roeder et al., 1970; Roeder, 1972).

B. THE HYPOTHESIS

Efficient transfer of mechanical energy from a light and compressible medium such as air to a dense and incompressible medium such as tissue fluid requires an impedance—or density-matching mechanism. A lever operating about a fulcrum provides an analogy; a small force applied to the long arm causes a large force that acts over a proportionally shorter distance at the short arm. Another analogy to the solution reached by some detectors of airborne sound is supplied by a windmill used to pump water. If this process is to be efficient air currents must act on a large area of sail in order to drive a pump of small stroke and bore.

One of the simplest acoustic systems in insects finds analogy in the lever. Extremely fine, long, and delicately poised hairs on the cerci of cockroaches and crickets are sensitive both to air currents and to low-pitched sounds (Pumphrey and Rawdon-Smith, 1936). Each hair articulates with a socket set in the cuticle and connects just below this "fulcrum" with the process of a sense cell. The arm lengths of this "lever" are certainly greater than

FIG. 8. The white-lined sphinx (*Celerio lineata*) approaching (A) and beginning to suck nectar (B) from a blossom of white phlox. In A the proboscis is partially extended but barely visible in the shadow cast by the wings; in B it is fully uncoiled and partially inserted into the blossom. The two frames do not belong to a single sequence. The single photos were made in dim light but the noise and light of the electronic flash caused the moth to dart away after each picture was taken, only to return and resume feeding at the same flower within a few seconds. The photos were taken of moths in captivity. (From Roeder and Treat, 1970.)

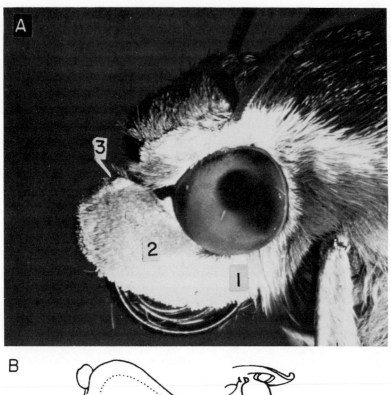

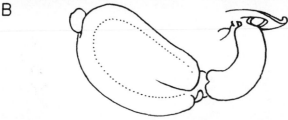

FIG. 9. A, Lateral view of the head of the white-lined sphinx. The large labial palp, its lateral surface covered by scales, extended forward from the articulation of its proximal segment (1) with the cranial floor ventral to the compound eye. The bulbous second segment (2) conceals the distal lobe of the pilifer and most of the coiled proboscis. The distal segment of the palp (3) is small and of unknown sensory function. B, Outline diagram of the labial palp denuded of scales and approximately to the scale of A. The articulation of the proximal segment with the cranium is freely movable and under muscular control. (A, from Roeder *et al.*, 1968; B, from Roeder and Treat, 1970.) A, Copyright 1968 by the American Association for the Advancement of Science.

100:1 so that large displacements of the external shaft cause minute displacements at the sense cell—probably as small as 0.5 Ångstrom (Pumphrey, 1940).

The eardrum and middle ear of higher vertebrates and the various tympanic organs of insects operate as impedance-matching mechanisms more in the style of the windmill water pump. A thin membrane of relatively large area and backed by an air space is coupled near its center to a column of fluid relatively small in cross section. This contains the sensing cells. Air pressures or displacements acting on the light membrane are thus "geared down" to produce large forces of small amplitude at the receptor surface.

With these examples in mind it was postulated that the palp of the white-lined sphinx operates on the lever principle. Its shape indeed suggests this. The segmented palps of moths in this subfamily are especially thin-walled and bulbous. They are freely articulated to the cranium through a socket in its floor (Fig. 9b). The fact that the palp is occupied mostly by an air sac makes it buoyant and liable to angular displacement by air movements about its articulation with the cranium. It was thought that all that was needed to convert the palp into a detector of airborne vibrations was a mechanoreceptive sense cell located just within the cranium and connected to the base of the palp close to its articulation. Such a sense cell would then be subject to small angular displacements at a point near the "fulcrum" (socket) when sound waves caused larger displacements of the buoyant second segment of the palp that would thus serve as the long arm of the lever.

It is now somewhat depressing to recall the amount of time and energy spent by several people in seeking antomical and neurophysiological evidence in support of this hypothesis. Much of the evidence was equivocal and some was positive, but what was more important, a still smaller amount of the evidence was unequivocally negative. For instance, attempts to immobilize the palp by fixing its outer surface to the cranium by wax bridges invariably failed to reduce the acoustic sensitivity. This result was difficult to reconcile with the fact that a small lateral displacement of the palp about its articulation with the cranium reduced acoustic sensitivity 100-fold (Roeder and Treat, 1970).

The final explanation of these paradoxical findings came not through conscious cerebration but during what can only be described as displacement activity of the frustrated experimenter. I was fretfully scraping with a fine needle at the multitude of projections and other structures on the cranium of a moth lacking its palps while spikes from its acoustic interneuron issued from a loudspeaker.

C. The Distal Lobe of the Pilifer

The needle tip struck a minute fingerlike structure near the base of the moth's tongue and elicited a vigorous burst of interneuron spikes. This structure turned out to be the distal lobe of the pilifer, a derivative of the labrum, a structure completely new to me and, as it turned out, familiar only to a few taxonomic specialists. This discovery triggered a second burst of progress devoted to studying the anatomy of the distal lobe, its mechanical relation with the palp, and the physiological properties of this unusual ear (Roeder *et al.,* 1970; Roeder, 1972).

The distal lobe of the pilifer after removal of the palps is shown in Fig. 10. It is about 0.3 mm long and completely smooth on its convex lateral surface. When the palps are in their normal position on either side of the base of the moth's proboscis they completely overlie and hide the distal lobes from view. Under these conditions the external surface of the medial wall of the bulbous second palpal segment appears to be in contact with the lateral surface of the distal lobe.

The juxtaposition of these two appendage derivatives appears to provide the moth with an acoustic detector in the following manner. Sound waves striking the thin lateral and medial walls of the second palpal segment throw them into vibration. The medial wall is traversed at one point by a longitudinal fold or slight thickening that touches the distal lobe of the pilifer when the palps are held in their normal adducted position (Fig. 10). Vibrations of this fold are here transferred to the distal lobe, which contains the acoustic sensor.

This arrangement finds an analogy in the windmill water pump. The windmill sail is equivalent to the extensive, thin, medial wall of the second

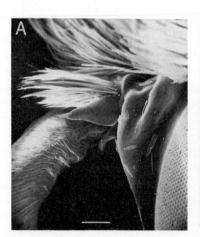

FIG. 10. Scanning electronmicrographs of the distal lobe of the pilifer of *Celerio lineata* exposed by removal of the labial palp. A, Lateral view. The roughly triangular distal lobe lies lateral to the base of the proboscis (extending downward to the left), and just to the left of the border of the compound eye. B, Enlarged view of the distal lobe. Its smooth lateral surface is normally in contact with the medial surface of the overlying second palpal segment. Scale in A, 0.1 mm. (From Roeder, 1972.)

palpal segment. Its displacements by sound are transferred via the fold to the relatively minute distal lobe, which is equivalent to the water pump.

Detailed evidence for this arrangement is presented elsewhere (Roeder, 1972). First, it is immediately apparent why amputation or deflection of the labial palp reduces acoustic sensitivity about 100-fold but does not wholly abolish it. The "sail" has been disconnected from the "pump" which, though still intact, now operates at greatly reduced efficiency. If an artificial sail consisting of a small "flag" of thin plastic film is mounted so that its tip touches the exposed distal lobe acoustic sensitivity is partially or almost completely restored, as measured by the interneuron response in the nerve cord. In another experiment the distal lobe was vibrated mechanically rather than acoustically by bringing it into contact with a fine glass probe connected to a quartz crystal oscillating at ultrasonic frequencies. The crystal produced deflections of known amplitude, making it possible to measure the sensitivity of the distal lobe to linear displacements. At ultrasonic frequencies the distal lobe was found to be sensitive to displacements as small as 0.3 Ångstrom.

The most novel characteristic of this insect ear is that its two components—the impedance-matching portion and the sensory portion—derive from appendages belonging to different segments of the head and merely brought into casual external contact. Passive deflection of the palp makes it possible to disconnect and then to reconnect these two components any number of times in the course of a single experiment. This offers an unusual opportunity to examine their separate physical contributions to sound detection. This reminds one of the permanent union of parts of head segments that took place during the evolution of the ear of terrestrial vertebrates. Here the spiracular gill cleft became associated with the membranous labyrinth and mechanically coupled to it first by the hyoid bone and then by the jaw articulation of an extinct reptile. A mammal deprived of its ear drums and auditory ossicles is in a condition roughly analogous to *Celerio* minus its palps.

Neurophysiologically, much remains to be studied. There appears to be only a single acoustic sense cell in the distal lobe, although it is still to be located and described. The sensory nerve running from the distal lobe to the moth's brain has been traced, but sensory or first-order impulses in it have been recorded only on a few occasions. The second-order or interneuronal response is easily registered and served as the criterion signal in all the experiments described above. Still to be discovered are the destinations of this spike signal and the manner in which it releases the evasive behavior observed at the jasmine hedge in Texas.

This brings the story back once more to questions of behavior and evolution. Do bats in fact attack the white-lined sphinx and its relatives as they hover and feed from blossoms? This event has not been reported although there is some positive cirumstantial evidence (Roeder and Treat, 1970).

Do other aerial predators of the moths perhaps betray their approach by sound? *Celerio* flies often in late dawn and early dusk and its palp–pilifer organ responds in the sonic as well as the ultrasonic range of sound intensities. Why, out of five subfamilies of hawkmoths, has the palp–pilifer combination evolved into a sensitive acoustic detector only in one subfamily (Choerocampinae, 4 species tested). Many hawkmoths belonging to other subfamilies are similar in size and feeding habits yet they fail to show a neurophysiological response to sound (22 species tested). These and similar questions will be answered only by field observations of predators and prey.

IX. Epilog

One of the methods of comparative biology is to define the differences between two species and at the same time to seek out their similarities. I have tried to do this with the two scientific "species," ethology and neurophysiology. My conclusion is that unlike the history of sibling animal species ethology and neurophysiology had divergent origins and now form an interbreeding population.

Some of their differences in origin and early objectives are outlined in Section II of this essay. Section III attempts to summarize the major concepts of neurophysiology. It may even seem to widen the gap since it suffers from the defects common to generalizations—some readers will find it inadequate and others incomprehensible. However, I hope that a few readers will find in the second part of this section common ground in the *terra incognita* of both fields. Section IV lists and examines a few of the misconceptions of one field about the other. Some of these misconceptions are either emphemeral or they arise from traditional or methodological differences. In other cases, I have attempted to show that misconceptions such as pertain to objectivity and precision are common to both fields.

So much for conceptual differences that will always be subjects for armchair discussion. How and where do ethology and neurophysiology find congruence in the field and laboratory? At what point do they become common in the real business of dealing with the actions of animals? In Section V, "Inside the Skin," I have examined in some detail what seem to me to be an operational but very minor interface, so minor that whether animal behavior is scrutinized from a neurophysiological or from an ethological viewpoint the boundary of the "skin" must be crossed and recrossed numerous times if research on animal behavior and its mechanisms is to give real intellectual satisfaction. At least, I have found this to be so, and the sections following Section V are intended to illustrate it. These are a personal account of matters that have interested me in insects, and I would not and could not separate them into ethological and neurophysiological components. I can only conclude that life would have been a great deal less interesting, in more senses than one, if I had felt constrained to be a specialist in either field.

References

Adams, W. B. 1971. Intensity characteristics of the noctuid acoustic receptor. *J. Gen. Physiol.* **58**, 562–579.

Adrian, E. D., 1930. The activity of the nervous system of a caterpillar. *J. Physiol. (London)* **70**, 34–35.

Adrian, E. D. 1931. Potential changes in the isolated nervous system of *Dytiscus marginalis. J. Physiol. (London)* **72**, 132.

Bonsdorff, G. 1709. Fabrica usus, et differentiae antennarum in insectis. Dissertation (O. B. Rosenstein, respondent) Abo, Finland. [Transl. by Sharp, J. 1833. *Field Natur.* **1**, 292.]

Bullock, T. H., and Horridge, G. A. 1965. "Structure and Function of the Nervous Systems of Invertebrates." Freeman, San Francisco, California.

Chadwick, L. E. 1953. The motion of the wings. *In* "Insect Physiology" (K. D. Roeder, ed.), pp. 578–614. Wiley, New York.

Cohen, M. J. 1965. The dual role of sensory systems: detection and setting central excitability. *Cold Spring Harbor Symp. Quant. Biol.* **30**, 587–599.

Davis, W. J., and Kennedy, D. 1972. Command neurons controlling swimmeret movements in the lobster. *J. Neurophysiol.* **35**, 1–29.

Eccles, J. C. 1957. "The Physiology of Nerve Cells." Johns Hopkins Press, Baltimore, Maryland.

Eggers, F. 1928. "Die stiftfuhrenden Sinnesorgane." Zool. Baustein, Berlin.

Ghiradella, H. 1971. Fine structure of the moth ear. 1 The transducer area and connections of the tympanic membrane in *Feltia subgothica* Haworth. *J. Morphol.* **134**, 21–45.

Griffin, D. R. 1958. "Listening in the Dark." Yale Univ. Press, New Haven, Connecticut.

Hodgkin, A. L. 1951. The ionic basis of electrical activity in nerve and muscle. *Biol. Rev. Cambridge Phil. Soc.* **26**, 339–409.

Hodgkin, A. L. 1964. "The Conduction of the Nervous Impluse." Liverpool Univ. Press, Liverpool.

Hubel, D. H., and Wiesel, T. N. 1965. Receptive fields and functional architecture in the non-striate visual areas (18 and 19) of the cat. *J. Neurophysiol.* **18**, 229–289.

Kalmijn, A. T. 1971. The electric sense of sharks and rays. *J. Exp. Biol.* **55**, 371–383.

Kammer, A. 1968. Motor patterns during flight and warm-up in lepidoptera. *J. Exp. Biol.* **48**, 89–109.

Katz, B. 1966. "Nerve, Muscle, and Synapse." McGraw-Hill, New York.

Lettvin, J. W., Maturana, H. R., McCullock, W. S., and Pitts, W. H. 1959. What the frog's eye tells the frog's brain. *Proc. IRE* **47**, 1940–1951.

Lissman, H. W. 1958. On the structure and function of electric organs in fish. *J. Exp. Biol.* **35**, 156–191.

Lorenz, K. 1937. Über die Bildung des Instinktbegriffes. *Naturwissenshaften* **25**, 289–331.

M'Cracken, M. 1907. The egg-laying apparatus in the silkworm (*Bombyx mori*) as a reflex apparatus. *J. Comp. Neurol.* **17**, 262.

Milburn, N., Weiant, E. A., and Roeder, K. D. 1960. The release of efferent nerve activity in the roach, *Periplaneta americana,* by extracts of the corpus cardiacum. *Biol. Bull.* **118**, 111–119.

Pringle, J. W. S. 1957. "Insect Flight." Cambridge Univ. Press, London and New York.

Prosser, C. L., and Brown, F. A. 1961. "Comparative Animal Physiology." Saunders, Philadelphia, Pennsylvania.

Pumphrey, R. J. 1940. Hearing in insects. *Biol. Rev. Cambridge Phil. Soc.* **15**, 107–132.

Pumphrey, R. J., and Rawdon-Smith, A. F. 1936. Synchronized action potentials in the cercal nerve of the cockroach, (*Periplaneta americana*). *J. Physiol. (London)* **87**, 4–5.

Roeder, K. D. 1935. An experimental analysis of the sexual behavior in the praying mantis. *Biol. Bull.* **69**, 203–220.

Roeder, K. D. 1955. Spontaneous activity and behavior. *Sci. Mon.* **80**, 262–270.

Roeder, K. D. 1964. Aspects of the noctuid tympanic nerve response having significance in the avoidance of bats. *J. Insect Physiol.* **10**, 529–546.

Roeder, K. D. 1966a. Acoustic sensitivity of the noctuid tympanic organ and its range for the cries of bats. *J. Insect Physiol.* **12**, 843–859.

Roeder, K. D. 1966b. Interneurons of the thoracic nerve cord activated by tympanic nerve fibres in noctuid moths. *J. Insect Physiol.* **12**, 1227–1264.

Roeder, K. D. 1967a. "Nerve Cells and Insect Behavior," 2nd Ed. Harvard Univ. Press, Cambridge, Massachusetts.

Roeder, K. D. 1967b. Turning tendency of moths exposed to ultrasound while in stationary flight. *J. Insect Physiol.* **13**, 873–888.

Roeder, K. D. 1970. Episodes in insect brains. *Amer. Sci.* **58**, 378–389.

Roeder, K. D. 1971. Insect flight behavior: some neurophysiological indications of its control. *In* "Progress in Physiological Psychology" (E. Stellar and J. M. Sprague, eds.), Vol. 4, pp. 1–36. Academic Press, New York.

Roeder, K. D. 1972. Acoustic and mechanical sensitivity of the distal lobe of the pilifer in choerocampine hawkmoths. *J. Insect Physiol.* **18**, 1249–1264.

Roeder, K. D., and Treat, A. E. 1957. Ultrasonic reception by the tympanic organ of noctuid moths. *J. Exp. Zool.* **134**, 127–158.

Roeder, K. D., and Treat, A. E. 1970. An acoustic sense in some hawkmoths (Choerocampianae). *J. Insect Physiol.* **16**, 1069–1086.

Roeder, K. D., Tozian, L., and Weiant, E. A. 1960. Endogenous nerve activity and behavior in the mantis and cockroach. *J. Insect Physiol.* **4**, 45–62.

Roeder, K. D., Treat, A. E., and Vande Berg, J. S. 1968. Auditory sense in certain hawkmoths. *Science* **159**, 331–333.

Roeder, K. D., Treat, A. E., and Vande Berg, J. S. 1970. Distal lobe of the pilifer: an ultrasonic receptor in choerocampine hawkmoths. *Science* **170**, 1098–1099.

Sherrington, C. S. 1906. "The Integrative Action of the Nervous System." Yale Univ. Press, New Haven, Connecticut. (Reissued as a Yale Paperbound, 1961.)

von Holst, E. 1935. Die Koordination der Bewegung bei den Arthropoden in Abhängigkeit von zentralen und peripheren Bedingungen. *Biol. Rev. Cambridge Phil. Soc.* **10**, 234–261.

Wilson, D. M. 1961. The central control of flight in a locust. *J. Exp. Biol.* **38**, 471–490.

Wilson, D. M. 1967. An approach to the problems of rhythmic behavior. *In* "Invertebrate Nervous Systems" (C. A. G. Wiersma, ed.), pp. 219–230. Univ. of Chicago Press, Chicago, Illinois.

The Orientational and Navigational Basis of Homing in Birds

WILLIAM T. KEETON

SECTION OF NEUROBIOLOGY AND BEHAVIOR
CORNELL UNIVERSITY
ITHACA, NEW YORK

The last few years have witnessed a marked acceleration of research in avian orientation and navigation, with the result that there have been many important changes in our thinking about this ever-intriguing and still poorly understood phenomenon. Consequently, it seems appropriate at this time to take stock of one aspect of the subject—homing—even though three reviews (Schmidt-Koenig, 1965; Wallraff, 1967; Gwinner, 1971) and an excellent theoretical paper on avian orientation (Griffin, 1969) are only a few years old. Because those publications, considered together, provide a clear picture of the research done prior to the late 1960's, and of the ideas current at that time, I shall make no attempt here to go over all the same ground.

Instead, my focus will be on those aspects of homing research that have been most under scrutiny in the last 5 years or that are, in my judgment, likely to figure prominently in the work of the next five. I admit at the outset that this will be a somewhat biased treatment, in that I shall freely express my personal views on topics that are controversial, and shall, moreover, draw heavily on experimental results from my own research group—results that, in some cases, are published here for the first time.[1]

Though this paper will be concerned chiefly with homing, and hence to a large extent with research on pigeons, work on migratory orientation will be treated whenever it bears importantly on the topics discussed. A paper complementary to this one—i.e., emphasizing migratory orientation but discussing homing when appropriate—has been prepared concurrently by my colleague, Stephen T. Emlen (in preparation), and readers are referred to it. The most recent book-length treatment of the many aspects of avian migration is by Schüz et al. (1971).

I. HOMING CAPABILITIES OF BIRDS

Before we begin a detailed consideration of experiments designed to probe the mechanisms of homeward navigation, a brief summary of the homing feats birds can actually perform is in order.

Certainly the oft-cited return of a Manx shearwater from Boston, Massachusetts, to its burrow on the island of Skokholm (off the southwest corner of Wales), a distance of more than 3,000 miles negotiated in only 12½ days (Matthews, 1953c), remains one of the most impressive examples of long-distance homing, as do the rapid returns to Skokholm of other Manx shearwaters released at various points in Britain and Europe (Lack and Lockley, 1938; Matthews, 1953c, 1964). Returns to Midway Island of 82% of the Laysan albatrosses released by Kenyon and Rice (1958) at distances ranging from 1,665 to 4,120 miles must also be listed among the most spectacular homing records known; one of these birds covered the 4,120 miles from Luzon to Midway in 32 days. Griffin (1940) and Billings (1968) had high percentages of returns of Leach's petrels released at distances of 135 to 2,980 miles. Experiments with other procellariiform species would doubtless yield similar results.

Respectably rapid homing from long distance by a high percentage of the individuals released has also been found in gannets (Griffin and Hock,

---

[1] This review is based on the published literature through mid-1972, except that some material then in press or in preparation, for which I was able to read manuscripts, is included. Footnote references to some of the most important papers published between mid-1972 and November, 1973, have been added.

1949), white storks (Wodzicki *et al.*, 1938, 1939), herring gulls (Griffin, 1943), lesser black-backed gulls (Matthews, 1952), various species of terns (Watson and Lashley, 1915; Dircksen, 1932; Griffin, 1943), various species of swallows (Rüppel, 1934, 1936, 1937; Wodzicki and Wojtusiak, 1934; Southern, 1959, 1968), swifts (Spaepen and Dachy, 1952, 1953; Spaepen and Fragniere, 1952; Schifferli, 1942, 1951), and starlings (Rüppell, 1935, 1936, 1937; Birner *et al.*, 1968).

It is apparent that the species mentioned in the above two paragraphs are ones in which the adults range rather widely while foraging. Homing experiments have been performed on a variety of species that limit their foraging during the breeding season to a very restricted territory (see listing in Matthews, 1955a), and in a few cases individual birds have returned from remarkably long distances. But in general the percentages of returns have been low and the speeds slow, even when the releases were only a few miles from the nests. This does not necessarily mean that such species lack the ability to orient toward a goal in other biological contexts—it seems quite possible, in fact, that some of them perform goal orientation when they are nearing the end of their migratory flight, and a few researchers (e.g., Rabøl) think they use goal orientation throughout the migratory voyage. But it does suggest that experimenters wishing to use breeding birds in homing tests are well advised to choose a species that does not confine its activities to a very small territory.

Though much can be learned from experiments with wild birds that exhibit good homing behavior, such studies are plagued by many technical problems: (1) experiments can usually be conducted during only a brief part of the year; (2) the birds are often hard to catch and it is difficult to assess their motivational state; (3) little is usually known about the birds' previous experience (except when fledglings are used, and they seldom home well); (4) many species are difficult to maintain in captivity long enough to perform manipulations on them (e.g., clock-shifts) without serious alterations of their physical health or motivational condition; (5) it is frequently very difficult to determine exactly when a bird arrived home in order to compute its speed accurately or even, in some cases, to tell whether it got home at all; and (6) it is seldom possible to perform many tests with the same individuals. It is not surprising, therefore, that the bulk of the research on the mechanisms of homing has been performed on pigeons.

Modern-day homing pigeons are the products of centuries of intense selection for the sort of behavior we are here considering (Levi, 1963), and their use makes it possible to avoid most of the problems just enumerated. It is important to understand, however, that really outstanding performances can be expected from homing pigeons only if much care is given to the details of their management (composition of the feed used, schedule and amount of feeding, exercise, maintenance of proper motivation to home, disease prevention, etc.). Indeed, it is often their skill in attending to these details that distinguishes the fanciers whose birds are consistent winners in races (Aerts, 1969; Barker, 1958; Whitney, 1969; Petit and Depauw, 1952; Hutton, 1964, 1966; Violette, 1958). Scientists who ignore such considerations do so to their own detriment. Homing pigeons were developed for performance under certain rather specific conditions, and one cannot profit fully from their breeding if those conditions are not taken into account.

There are many different inbred strains of homing pigeons, and fanciers debate their respective merits at great length in the journals of the racing sport, but for our purposes we can sort them into two groups: short-distance strains (which excel in races of 100–300 miles) and long-distance strains (which excel in races of 400–600 miles or more). Fanciers who are successful breeders and trainers of long-distance strains expect their birds to return from races of 500 or 600 airline miles the same day they are released (provided the weather is reasonable), i.e., the birds are released shortly after sunrise and they are expected to be home before sunset that day. I personally knew a fancier in Boston whose birds regularly completed the 600-mile race in a day, year after year. For example, in 1969 he entered 14 birds in the 600-mile race; 12 were home the same day and the other 2 returned early the next morning. A club in Fort Wayne, Indiana, has conducted a yearly 1,000-mile race since 1958 and has frequently had returns in less than 3 days, even though the birds have to stop for the night and must take time to search for food and water (Snyder, 1971); one fancier entered a total of 35 birds in these 1,000-mile races over the course of 8 years, and lost only 6 (Haffner, 1966). Review of the records of hundreds of pigeon races leads me to the conclusion that the limitation on the distances from which pigeons can successfully home is primarily a function of the physical hardships involved rather than of orientation capabilities. This conclusion receives some support from the recent transoceanic displacement experiments conducted by Wallraff and Graue (1973).

II. SOME EARLY MODELS OF HOMING

A. TYPES OF ORIENTATION

In 1955, Griffin distinguished three different types of orientational ability. Type I, usually called piloting, is steering a course on the basis of familiar landmarks. Type II is the ability to head in a given compass direction without reference to landmarks. Type III, the most complex, is true navigation, which is the ability to orient toward a goal (e.g., home), regardless of its direction, by means other than recognition of landmarks (Griffin, 1955). It should be understood at the outset that all three types of orientation prob· ably occur among birds, but what we are most concerned with here is the possible role each may play in homing.

That simple pilotage cannot fully explain pigeon homing can be established easily. When released at distant sites, young pigeons that have never previously been taken away from their loft usually depart nonrandomly in a roughly homeward direction; sometimes their bearings are already homeward directed after only 40 seconds (Fig. 1), which indicates clearly that they do not merely fly a search pattern until they encounter familiar territory (Wallraff, 1970b). Even pigeons that have been confined in aviaries all their lives, with no opportunity to become familiar with any landmarks not visible from the aviaries, show a marked tendency to travel toward home upon displacement and release, though they may not actually arrive at their destination (Fig. 2) (Kramer, 1959; Wallraff, 1966b, 1970b). There are many other reasons for concluding that landmarks play a decidedly secondary role in pigeon homing, but I shall defer discussion of them until later.

It is also easy to establish that simple compass orientation is not a sufficient explanation of pigeon homing. Figure 1 can be cited again in this context, as it shows that inexperienced birds released in different directions from their home loft do not take up some standard bearing without regard for the position of the release point relative to home; instead, their choice of

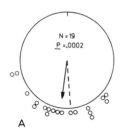
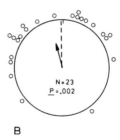

Fig. 1. Bearings of first-flight pigeons 40 seconds after release at sites 63 miles north of the loft (A), and 65 miles south of the loft (B). (Adapted from Wallraff, 1970b.) [In this and all later figures, the following format will hold: A small line at the top of the large circle indicates geographic north. A dashed line shows the home direction. Each symbol on the periphery of the large circle indicates the bearing of one bird; the mean of these bearings is shown as an arrow whose length (r) is inversely proportioned to the extent of scatter (i.e., a circularly uniform distribution would result in a mean vector of length 0, whereas a distribution with no scatter—all bearings the same—would result in a mean vector of length 1, which would be shown as an arrow that reached the periphery of the large circle). The size of the sample (N) and the uniform probability (P) of the bearings under the Rayleigh test are shown.]

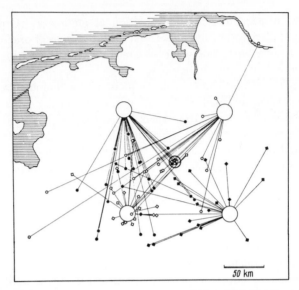

Fig. 2. Recovery points of aviary pigeons released at four sites (circles) 90 km from Osnabrück, West Germany (smaller central circle). The lines connecting the release sites with the recovery points show a clear homeward tendency. (From Wallraff, 1970b.)

direction is quite evidently goal directed. The same conclusion must be drawn when birds from different lofts, released simultaneously at the same unfamiliar site, depart in their respective homeward directions (Fig. 3).

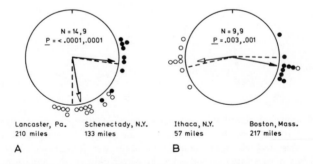

Fig. 3. Two tests in which pigeons from different lofts were released on the same day at the same site. The birds from each loft showed good homeward orientation. (From Keeton, unpublished data). [Whenever, in this and later figures, two sets of bearings are included in the same drawing, one set (the controls in the case of experimental procedures) will be shown as open symbols, and the other set (experimentals) as solid symbols; open or solid arrow heads distinguish the respective mean vectors. The first-given values of N and P refer to the open bearings and the second values to the solid bearings.]

Having seen, then, that neither Griffin's Type I nor Type II orientation suffices to explain pigeon homing, we must conclude that Type III, true navigation, is involved, and we must examine possible explanations for such navigation. Let me state at the outset that, in my opinion, no satisfactory overall hypothesis exists at present. The old hypotheses have been disproved and no convincing new ones have been proposed. Nonetheless, many new pieces of the puzzle have been discovered in recent years and we can hope that, with the discovery of a few more, a pattern for fitting them together will begin to emerge.

As a background for understanding the newer work, let us look briefly at the principal models proposed in the years 1947–1960, and at the evidence against them. More particularly, we shall single out the proposals of Yeagley, Matthews, and Kramer.

B. YEAGLEY'S HYPOTHESIS

The possibility that the earth's magnetic field might provide directional information to an orienting bird has been suggested off and on since 1882 (e.g., Viguier, 1882; Thauzies, 1898; Casamajor, 1926; Stresemann, 1935; Daanje, 1936, 1941). However, in the absence of any convincing supportive evidence, relatively little attention was paid to this possibility until Yeagley (1947, 1951) combined it with possible detection of the Coriolis force resulting from the earth's rotation (as suggested by Ising, 1945) to construct a possible bicoordinate model of avian orientation. According to Yeagley, the flying bird could detect the geographic variations in both the vertical component of the magnetic field and the strength of the Coriolis force. Since the roughly circular lines of equal magnetic strength are centered on the magnetic poles, and the lines of equal Coriolis force are similarly centered on the geographic poles, and since the magnetic and geographic poles are approximately 1,600 miles apart, the two sets of lines form a grid that could, in principle, provide a basis for navigation.

In a test of part of his hypothesis, Yeagley attached magnets to the wings of one group of birds and copper plates to the wings of another group. He reported (Yeagley, 1947) that the performance of the birds carrying the magnets was significantly poorer, but he later (Yeagley, 1951) admitted that in another test of this same sort no difference between the experimental and control birds was evident. Other investigators, attempting to repeat Yeagley's experiments, obtained uniformly negative results (Gordon, 1948; Matthews, 1951b; van Riper and Kalmbach, 1952).

Yeagley also performed tests in which birds from his Pennsylvania loft were released near a conjugate point (one having a similar combination of magnetic and Coriolis values) in Nebraska, and he reported that the birds oriented toward that point.[2] However, reanalysis of his data failed to support his claims (Thorpe, 1949; Griffin, 1952a; Matthews, 1951a, 1955a). His hypothesis has been criticized on theoretical grounds by a number of writers (e.g., de Vries, 1948; Slepian, 1948; Davis, 1948; Varian, 1948). The extremely negative reception accorded Yeagley's ideas had the effect not only of casting his hypothesis aside but also of discouraging further consideration of the possible use of magnetic cues in orientation.

C. Matthews' Hypothesis

A hypothesis of complete navigation by the sun alone was tentatively suggested by Matthews in 1951 and later expounded in considerable detail (Matthews, 1953a, 1955a). Matthews suggested that a displaced bird could determine its latitudinal displacement (and also geographic south) by observing the sun's movement in arc and extrapolating that arc to its noon position; comparison of the sun's noon altitude at the release site with its noon altitude at home (remembered) would indicate whether the bird was north or south of home, and how far. Longitudinal displacement could be determined by, in effect, comparing local sun time at the release site (determined by observing the sun's position on its arc) with home time as indicated by the bird's internal clock. By vectorially combining the latitudinal and longitudinal information thus obtained, the bird could determine the direction it must fly to get home.

Matthews' sun-arc hypothesis was the stimulus for much of the research on homing performed in the last two decades, and as such it must be reckoned one of the milestones in the history of this field. But unfortunately experiments by many investigators have yielded results contrary to those predicted by the hypothesis, and consequently it has been rejected by most scientists now actively engaged in research on pigeon homing. I know of no convincing evidence for the existence of sun navigation, as distinct from sun-compass orientation, in any bird species, though Bellrose (1972) has recently accepted (on what I consider very weak grounds) the idea that some "navigationally advanced" species may possess such a capacity. I haven't space here to discuss all the evidence, but shall mention those types that seem particularly compelling.

[2] Dr. Yeagley informed me (personal communication) that, in an experiment not reported in his publications, he released two pigeons every 50 miles as he drove from State College, Pennsylvania, to the conjugate point in Nebraska. He said for the first half of the distance every bird vanished in the eastern half of the circle (i.e., roughly toward the home loft), and for the second half of the trip every bird vanished in the western semicircle (i.e., roughly toward the conjugate point).

1. Some Theoretical Considerations

Kramer (1957) and Pennycuick (1960) criticized the hypothesis on theoretical grounds: (1) A rapidly flying bird would encounter serious problems of parallax in using any fixed object, such as features of the landscape, to measure changes in the sun's position. (2) The sun's path across the sky is not circular, as originally assumed by Matthews, and hence it would be very difficult for a bird to construct it by extrapolation from the tiny segment observed during the first few seconds, or even minutes, after release. Matthews (1968) has attempted to counter these criticisms, but his arguments are not convincing.

In an attempt to get around these objections, Pennycuick (1960, 1961) proposed an alternative sun-navigation mechanism. He suggested that a displaced bird could get all the information required to navigate home by observing only the sun's altitude and the rate of change of that altitude; no extrapolation need be performed. Though this variant of the sun-navigation hypothesis does avoid some problems inherent in Matthews' version, it is controverted by the same experimental evidence (below) and hence will not be discussed separately.

2. Sensory Requirements

Matthews' hypothesis requires that the bird possess extraordinary acuity of vision and accuracy of timekeeping. Some recent research results cast doubt on whether these requirements are met. First, P. Blough (1971) has found that the oft-repeated statements that the pigeon's powers of visual acuity are excellent are probably overly generous, especially if one is concerned with distant vision (monocular) rather than very near vision (binocular). Contrary to popular notion, the pigeon's distant vision appears to be decidedly inferior to man's. That it may also be inferior to that of many other birds (e.g., starlings, hawks) is suggested by the fact that the pigeon fovea is much shallower (Tansley, 1965; Galifret, 1968; Blough, 1971). The values of 0.5°–1° found by Meyer (1964) and 3.4° found by McDonald (1968) in tests of pigeons' ability to discriminate angles in the presentation of an artificial sun cannot be taken at face value because McDonald (1972, 1973) later found that his birds were not measuring angular deviations from the "sun" directly but were, instead, responding to angular differences in the position of shadows which would have been close enough to the birds' eyes to be detected by near-vision procedures. The same thing may have been true in Meyer's experiments.

Second, Adler (1963) and Meyer (1966), using operant conditioning techniques, and Miselis and Walcott (1970), monitoring activity rhythms under various regimes, have found that the pigeon's time sense is not sufficiently accurate for sun-based bicoordinate navigation, though it would suffice for sun-compass orientation.

3.   Is There a Rigidly Stable Navigational Clock?

In an attempt to evaluate the role of the bird's internal clock in homing, Walcott and Michener (1971) disrupted the chronometers of 6 pigeons by exposing them for 9 days to a random light-dark schedule like that used earlier by Matthews (1953a). During the 9 days, the birds were also given 30% D_2O drinking water, which has been shown to slow circadian rhythms [Suter and Rawson, 1968, for deer mice; T. H. Snyder (unpublished), for pigeons]. After this frightful treatment, the birds were put in an outdoor wire aviary at our lofts in Ithaca, New York, approximately 265 miles due west of their loft in Cambridge, Massachusetts. Here the birds were given normal water and were permitted an unobstructed view of the sun for a period of time ample for their internal clocks to become entrained to Ithaca time. They were then released individually from a site (unfamiliar to them) between Ithaca and Cambridge. Their vanishing bearings (Fig. 4B) were oriented homeward, not toward Ithaca, and were no different from those of control birds. This is inconsistent with Matthews' hypothesis, unless one is prepared to assume that there is a navigational clock (completely distinct from the activity-rhythm sun-compass clock) set so rigidly on home time that it is resistant both to a random L:D regime and to D_2O treatment. The necessity of a shift-resistant clock for sun-arc navigation was pointed out earlier by Hoffmann (1965), but he felt that the evidence available against sun-arc navigation at that time made it unnecessary to assume the existence of such a clock; he has recently reaffirmed this opinion (Hoffmann, 1971).

Matthews (1968) has, in fact, adopted the view that there is a stable

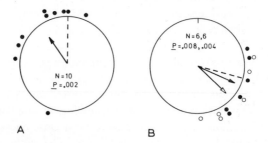

Fig. 4. Ten-mile bearings (radio) of pigeons subjected to mirror-box treatment (A), and random L:D plus D_2O treatment (B). In each case, the experimental birds show nonrandom, roughly homeward, orientations; in (B) there is no significant difference between the experimentals and the controls. (Adapted from Walcott and Michener, 1971.)

clock underlying pigeon navigation that is distinct from the sun-compass clock so many investigators have shown can easily be reentrained. Such a clock would, of course, be nearly essential if pigeons held at the release site for a long time are to be able to navigate home in the manner suggested by Matthews. That pigeons can indeed orient homeward after being held at the release site long enough for the clocks which we know they have to be entrained to local time is shown not only by such experiments as that of Walcott and Michener but also by the frequency with which adult pigeons, sold to new owners hundreds of miles away, return to their original home upon escape, even after several years have passed. No such stable clock is necessary if the birds use the sun only as a compass; indeed a bird released a very long distance east or west of home would, in principle, be able to use the sun compass more accurately in initial orientation if his internal clock were set to local time rather than to home time.

It is not uncommon for adult experienced pigeons to be resettled and raced at a new loft hundreds of miles from the original loft. I myself have resettled and flown such birds. Has the usually rigid navigational clock finally yielded to resetting in these cases? If so, then another difficulty arises: sometimes such resettled pigeons, after racing well to their new home, suddenly leave and return to their former home.[3] They should not be able to do this if their navigational clock is entrained to the sun time at the new home. Or do they have two navigational clocks, one for each home? Even more of a problem is the ability of mobile-loft pigeons, such as those used by the U.S. Army in World War II, to home to a succession of different locations (U.S. Army, 1923; U.S. War Department, 1945; Levi, 1963); when the lofts were moved 10 miles or less, the birds might, of course, find them by searching, but it is reported that birds accustomed to mobile lofts could be resettled after long moves in 8 to 10 days, and that birds based on warships could be resettled to new distant anchorages equally quickly. Must such pigeons have a separate rigid navigational clock for each new home locality, or is it not more reasonable to conclude that the only clock used in pigeon navigation is the sun-compass one for which so much evidence now exists? There is, to my knowledge, no shred of direct evidence in support of the rigid home-time clock for which Matthews argues.

4. The Sun's Altitude

Because Matthews' hypothesis requires that the birds determine north-south displacement from the sun's altitude, considerable experimental attention has been devoted to this subject.

Walcott and Michener (1971) confined 10 pigeons, one at a time, in a

[3] This happens most commonly when the bird's mate dies, or it suffers a severe fright, or is in some other way disturbed.

cage in a corner of the loft. The birds could see the outside through a window, the top part of which provided a mirrored view of the sun during the 3 hours (approximately) centered on noon. The mirrors were adjusted so as to decrease or increase the sun's noon altitude by 31'–70', which means the birds were seeing a sun appropriate to points 36–81 miles north or south of the actual location of the loft. After 10–14 days of such treatment, the birds were released at unfamiliar sites between the true home and the false home. They vanished toward home (Fig. 4A), despite their presumed incorrect memory of the sun's noon altitude at home.

A different approach to examining the effect of an incorrect memory of the sun's altitude at home was used by Matthews (1953a) himself. He confined 24 pigeons in a loft that permitted no view of the sun or sky for 6–9 days near the time of the autumnal equinox, when the day-to-day change in the sun's altitude is at its annual maximum. The birds were then released from a point 78 miles south of their home loft. The changes in the sun's declination during the days the birds could not see the sun were so great that the noon altitude of the sun at this southern release point on the days of the tests was actually less than it had been at home on the day the confinement began. Thus if the birds determine their latitudinal displacement by comparing the sun's noon altitude at the release site with its noon altitude at home when last seen, they should depart southward, away from home. This is precisely what Matthews reported his birds did (Fig. 5A).

Now, this was a very important experiment, because it provided the best (and nearly only) experimental evidence in direct support of Matthews' hypothesis, even though Matthews (1970) later sought to minimize its importance. It is therefore understandable that a number of investigators have tried to repeat the test (Rawson and Rawson, 1955; Kramer, 1955, 1957; Hoffmann, 1958; Keeton, 1970b). As Fig. 5 shows, none was successful; the birds always went north toward home. Though Matthews (1961, 1968) has attempted to explain away these negative results, it seems more reasonable to conclude that pigeons do not use the sun's altitude as a cue to latitude, or else that they are able to correct for seasonal changes even when they cannot see the sun. The first of these two possibilities is the more likely.

Matthews (1970) suggested that someone should perform the reciprocal experiment, i.e., prevent pigeons from seeing the sun for several days near the time of the vernal equinox and then release them north of the loft. We have, in fact, done this (Fig. 6), and again the results are contrary to Matthews' hypothesis. Moreover, since Matthews (1970) objected that our birds might simply be piloting by familiar landmarks,[4] we performed

----

[4] Matthews' conception of our proximity to the Adirondack and Catskill Mountains is rather faulty.

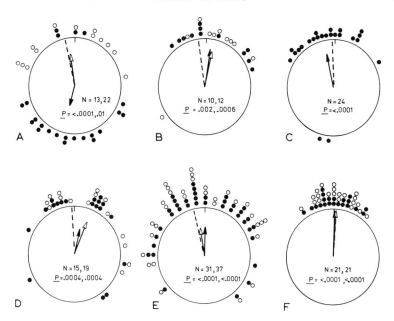

Fig. 5. Sun-occlusion experiments at the autumnal equinox. (A) Matthews' tests (1953a) appeared to show a separation of bearings, with the control birds departing northward, toward home, and the experimentals southward, as predicted by his hypothesis. But repeats of this experiment by (B) Rawson and Rawson (1955), (C) Kramer (1957), (D) Kramer (1955), (E) Hoffmann (1958), and (F) Keeton (1970b) all yielded negative results, i.e., both controls and experimentals departed northward. The reason for Matthews' atypical results is unknown.

one of these tests with first-flight youngsters (i.e., birds being taken away from home for the first time in their lives); the results remained the same (Fig. 6C).

In view of the negative results of both the mirror-box and sun-occlusion experiments, and of other more indirect evidence that the sun's altitude is not utilized (Hoffmann, 1954; Schmidt-Koenig, 1958), it has seemed time to cease debating the point and to turn our research energies into other more productive channels. However, just as the present paper was being completed, a report was published that reopens the question of whether the sun's altitude might play a role in avian orientation. Whiten (1972) reports that he has succeeded in conditioning pigeons mounted in a rotatable apparatus to indicate, by key pecking, whether they are pointed toward home or opposite to home, using the sun's altitude as their cue. There seems to be no question that his pigeons learned their task well, but one can question what that task was. Were the birds really using the home direction as a factor in determining their response? Or did they simply learn that the task

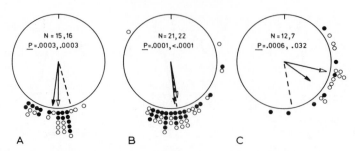

FIG. 6. Sun-occlusion experiments at the vernal equinox. (A) Experienced pigeons released 30.4 miles NNW of Ithaca, April 8, 1970. (B) Experienced birds released 43 miles N, March 29, 1972. (C) First-flight birds released 41 miles N, March 31, 1972 (normal unmanipulated first-flight pigeons regularly vanish to the southeast at this site). In all three tests, the bearings of the control and experimental birds were similar. (From Keeton, unpublished data.)

was to discriminate north from south by means of their sun compass and then to give a positive response when pointed north if the sun (or artificial sun) was high and a positive response when pointed south if the sun was low. In other words, there is not yet any assurance that the birds were orienting homeward. We must await further results before deciding the matter. In particular, it would be helpful to know the answers to the following two questions: (1) Do birds, conditioned at the north site only, show positive responses to the homeward direction when given unrewarded trials at the south site? (2) Do the pigeons learn just as easily to give positive responses to the coupling of a high sun with south pointing and a low sun with north pointing?

If it should turn out that Whiten's experiments merely show that pigeons can discriminate between two quite different sun altitudes when rewarded for doing so, as Meyer (1964) has already demonstrated, we would nonetheless know one new thing—that they can learn to couple this discrimination with a sun-compass north-south discrimination. Nevertheless, the weight of evidence at present indicates that this potential ability is not ordinarily used. in navigation.

5. Clock-Shift Experiments

Some of the most telling evidence against the Matthews hypothesis comes from experiments in which pigeons' internal clocks have been artificially shifted forward or backward. Such birds, upon release in a test of initial orientation, ordinarily depart in directions that deviate in a predictable way from the directions chosen by control birds. Thus, in the northern hemisphere, a 6-hours fast (clockwise) shift sends the birds roughly 90° counterclockwise from the controls, a 6-hours slow (counterclockwise) shift sends them roughly 90° clockwise from the controls, and a 12-hours shift sends them roughly 180° from the controls. In other words, a shift of a quarter of a day (6 hours) results in bearings deflected a quarter of a circle, etc. This was first demonstrated in extensive tests by Schmidt-Koenig (1958, 1960, 1961), and has since been confirmed by other investigators (e.g., Graue, 1963; Keeton, 1969); it remains difficult to explain why Matthews' (1955b) earlier attempts failed.

As Schmidt-Koenig (1965) has emphasized, it appears from such clock-shift tests that pigeons completely disregard such features of the sun's arc as altitude, rate of change of altitude, direction of movement (up or down), and respond only to its azimuth position. Thus the bearings chosen by clock-shifted birds are regularly consistent with use of the sun as a simple compass but not with use of the sun in the manner suggested by Matthews. For exam-ple, suppose we clock-shift pigeons 6 hours slow and then release them 100 miles due south of their home loft at local noon (Fig. 7). What might we expect of pigeons using Matthews' sun-arc navigation scheme? Let us dis-regard the matter of north-south displacement and consider only the east-west problem. If we assume that the clock we have manipulated is *the* clock used in homing (and we have seen that there is no evidence for any other), then according to Matthews' hypothesis the birds should consider their in-ternal time (6:00 a.m.) as home time and, upon observing a noon sun at the release site, conclude that they are approximately 4,000 miles east of home. Therefore the birds should depart almost due west. In fact, when one does this experiment, the birds vanish nearly due east. As we shall see later, this result can be explained if the birds are obtaining only compass

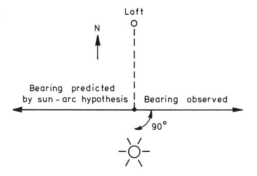

Fig. 7. Diagram of a noon release of 6-hours-slow clock-shifted pigeons at a site 100 miles south of their loft. The birds' internal clock indicates it is 6:00 a.m., so, if they are using sun-arc navigation, sight of a noon sun should lead them to determine their location as thousands of miles east of home, and they should depart westward. They actually depart eastward, as predicted by Kramer's map-and-compass model (see explanation in text, Section II,D).

information from the sun. Experiments of this type can be conducted with various combinations of shifts (forward, back) and directions (north, east, south, west), and the results continue consistent with use of the sun as a simple compass.

Matthews (1968) has criticized tests utilizing such massive shifts by pointing out that a bird might interpret the shift itself as a displacement. In this case, the birds in our example might have assumed that they were incarcerated at a point 4,000 miles west of home, which would explain their eastward departures, but Matthews appears to be inconsistent here. If a bird, upon incarceration, were to interpret the experimental shift as a displacement in longitude rather than as a change in his chronometer, and if he really responds to the sun's altitude, he should upon release detect that it is noon and hence conclude that he has suddenly undergone a second 4,000-mile displacement, this time eastward. Since the two displacements would cancel out each other, he might be expected to conclude that he is at the home longitude and, responding only to his southward displacement, depart northward, but he does not do this.

To get at this problem and at Pennycuick's (1961) objection that clock-shifted birds are ordinarily prevented from seeing the sun at home for so long they would no longer be able to remember its altitude by the time of the test, and at the same time to ascertain what sorts of information pigeons derive from the sun, one of my students and I (Alexander and Keeton, 1974) have recently clock-shifted birds in a room in their home loft while permitting them, for the 5–6 days of the shifting process, to sit in a wire aviary with a clear view of the sun and of the loft's surroundings during the overlap period between their shifted day and the true day. When released at test sites, such birds exhibited deflected bearings (Fig. 8) precisely like those obtained from birds held in light-tight rooms for the full

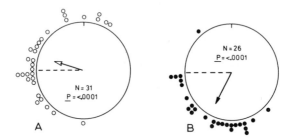

Fig. 8. Pooled bearings from three test releases of pigeons permitted a view of the sun at the loft while being clock-shifted 6 hours fast. The tests were under sun at a site 21 miles E of Ithaca, N.Y., June 24 and July 6 and 7, 1970. The difference between the means of the controls (A) and the experimentals (B) is 79°. (Adapted from Alexander and Keeton, 1974.)

shifting period. We conclude therefore that a clock-shift has its usual effect on initial orientation even when the birds have every evidence that their incarceration is at home and have seen the sun at home every day. We also conclude that while they are sitting in the aviaries the birds get no navigationally useful information from the sun other than that the light is on.

Walcott and Michener (1971) have subjected pigeons to very small clock shifts (5, 10, 15–20, and 120 minutes), which would be expected to have a much greater effect if the sun provides information for bicoordinate navigation than if it simply provides compass information. For example, if a bird used the sun only as a compass, at 45° latitude a time error of 20 minutes would cause him to deviate from the correct course by only about 5°, which would be too small a value for the available experimental procedures to detect, whereas if the bird used the sun for bicoordinate navigation he would be expected to head for a goal more than 240 miles east or west of home. Walcott and Michener found that such shifts caused only slight deflections of vanishing bearings. Hence they conclude that their experiments "argue strongly against the sun-navigation hypothesis." The same conclusion has been reached by Schmidt-Koenig (1972) after tests utilizing clock shifts of 30 and 120 minutes.

6. Orientation under Overcast

Recent evidence (Keeton, 1969, and Fig. 10 this paper; Baldaccini *et al.*, 1971; Wagner, 1972; Walcott, personal communication) that pigeons accustomed to flying in inclement weather can orient homeward quite accurately from distant release sites when the sun is completely hidden argues, at the very least, that no hypothesis such as Matthews' (or Pennycuick's) that predicates an essential role for the sun can possibly fully explain pigeon homing. I shall examine this topic more fully below (see Section II,D,2).

D. KRAMER'S MAP-AND-COMPASS MODEL

Kramer and his colleagues carried out most of the early work on orientation in circular cages, in which they showed that many birds, including pigeons, can use the sun as a compass (Kramer, 1950b, 1951, 1952, 1953b; Kramer and von St. Paul, 1950; Hoffmann, 1953). It is not surprising therefore that it was Kramer (1953b) who first suggested a way in which the sun compass could be utilized in homing. Now, a moment's thought suffices to convince that a compass alone is not sufficient for homeward navigation. Thus if a man were put out in an unfamiliar location far from home, given only a compass, and told to start walking toward home, he couldn't do it. Knowing where north, south, east, and west are is of no help if he doesn't know the home direction. But if a man has a map that tells him where he is, then a compass can be very helpful indeed. Kramer suggested that birds, too, must have "map" information to use in conjunction with their compass. According to his model, the bird first performs a map step, i.e., it determines its geographic position relative to home and also the theoretical home direction, and then it performs a compass step, i.e., it uses the sun compass to locate the deduced direction in the field.

Because Kramer never put forward an explicit description of what the map component is, his model is not a complete explanation of homing orientation. However, it has proved useful in describing what appears to happen when pigeons choose initial bearings at a release site. For example, in the release illustrated by Fig. 7, we saw that pigeons whose home loft was due north vanished eastward. We can explain this if we assume that the birds have some unspecified map information that tells them home is north, and that they try to use the sun compass to locate north. Their shifted internal clocks tell them it is 6:00 a.m., when the sun should be in the east. Hence north should be 90° counterclockwise from the sun, and it is this bearing that the birds choose. But it is actually 12:00 noon instead of 6:00 a.m., and the sun is therefore in the south, not in the east. This means that the birds' choice of a bearing 90° counterclockwise from the sun sends them east instead of north. Kramer's map-and-compass model is similarly effective in describing the results of tests utilizing other directions and other clock shifts.

Since the sun compass was virtually the only element in the pigeon navigation system that had been unequivocally demonstrated, and since its coupling with Kramer's hypothetical map had proved so useful in describing experimental results, many people came to assume that the sun was essential to pigeon homing, except where landmarks could provide a basis for piloting. There were, however, a few indications that this might not be so. I shall mention two here—nocturnal homing and homing under heavy overcast.

1. Nocturnal Homing

Though pigeons are normally diurnal animals and are nearly completely inactive at night, the Army Signal Corps demonstrated that some pigeons can be trained, albeit with great difficulty, to home at night (U.S. Army, 1923; Clarke, 1933; Levi, 1963). They had reasonable success with flights up to 30 miles and more limited success up to 50 miles. Moreover, during the years 1947–1950, a club in Hawaii conducted night races from distances up to 150 miles; some of these involved crossing stretches of open water up to 30 miles wide (Hitchcock, 1955; Levi, 1963).

von St. Paul (1962) reported success in experimental releases at night up to 12 miles, and she mentioned, but without giving data, having had returns from 25 and 43 miles. She has now graciously permitted my student, L. Goodloe, and me to examine the data from these tests. In six tests at 25–27 miles, an average of 19% of the birds returned during the night of the release. In a single test at 43 miles only 2 out of 19 birds returned.

I began tests of night homing in 1967, and Goodloe has continued these. We have had good returns from 20 miles and more limited returns from 30 miles (Keeton, 1970a). Radio tracking often reveals extraordinarily accurate bearings (Fig. 9), and the speeds are sometimes as great as 50 mph (Goodloe, 1974). The birds are probably not using a star compass, as some nocturnal migrants do, because they perform well under heavy (but high) overcast, and whether or not there is a moon. Unfortunately, however, we have not been able to get returns from great enough distances to rule out use of familiar light patterns as landmarks; certainly horizon glows of cities and beacon lights of airports are visible to human eyes from far away. Just how much flying pigeons can see at night is not clear, though we know their dark-adapted sensitivity is considerably less than that of humans (Blough, 1956).

We have not yet determined whether our lack of success in getting pigeons to home regularly at night from 50 miles or more is due to their lack of motivation to keep flying or to their inability to navigate. Consequently, we must conclude that night homing cannot yet be cited as convincing evidence against the necessity of the sun in pigeon navigation from unfamiliar release sites.

2. Homing under Heavy Overcast

Though the literature is full of reports that pigeons released under heavy overcast are disoriented, there have nonetheless been enough instances of homeward orientation under such conditions to cast doubt on the assumption that the sun compass is a necessity for navigation from unfamiliar sites. Thus Wallraff (1960a) found some "exceptions" in which vanishing bearings under overcast were not significantly different from bearings obtained at

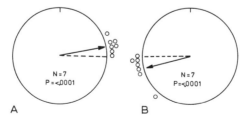

FIG. 9. Departure bearings (radio) of night-homing pigeons from release sites 21 miles west of Ithaca, N.Y. (A) and 21 miles east (B), August 16, 1971 and November 7, 1970. (From Goodloe, 1974.)

the same sites on sunny days. Moreover, reanalysis of Matthews' oft-cited data, using modern statistical procedures, reveals significant homeward orientation in a substantial proportion of the cases where randomness was originally claimed (Wallraff, 1966a; Keeton, 1969). For example, in 1953 Matthews presented the results of five "random" releases under overcast, but reanalysis under the Rayleigh test for uniformity on a circle (Batschelet, 1965, 1972) reveals that only two are actually random. The confused nature of the evidence led Schmidt-Koenig (1965) to state that, "The often repeated claim that orientation breaks down under overcast appears unwarranted so far."

It was against this background that our research on pigeon homing began at Cornell. Now, it is not unusual for Ithaca, New York, to experience prolonged periods of total overcast in spring and late fall, and it soon became apparent that our birds could orient homeward under such conditions quite accurately if they were first given minimal training. If the birds were forced to take their regular daily exercise flight at the loft regardless of the weather, and if they were given several short-distance (3–10 miles) training flights under heavy overcast or even in steady light rain, they could then be jumped to distant release sites with good results. The crucial thing seemed to be getting the pigeons accustomed to flying in inclement weather.

Having established to our satisfaction that our birds could orient accurately under total overcast, we set out to determine what sort of orientational mechanisms the birds were using under such conditions. A likely procedure seemed to be use of clock-shift tests under overcast. Schmidt-Koenig (1958) had previously performed a few "accidental" tests of this type but had stated (Schmidt-Koenig, 1965) that his results were "too few for a firm conclusion." Our first clock-shift tests (6 hours fast) under overcast were performed from 21 miles east of the loft (Keeton, 1969). Both the controls and the experimentals were homeward oriented and there was no significant difference between the two sets of bearings. Yet when the very same birds were again clock-shifted and released at the same site under sun, the results were dramatically different—the experimental birds chose bearings deviating approximately 90° from those of the control birds, as would be predicted if they were using the sun compass. Thus the following conclusions seemed warranted: (1) Our pigeons did use the sun as a compass when it was visible, as Schmidt-Koenig (1958, 1960, 1961) had earlier established for his birds. (2) When the sun was not visible, our birds could still orient toward home. (3) The homeward orientation under overcast was not due to some special ability of the birds to detect the sun's position when we could not, because if the sun's position were being used in the usual compass procedure then the shifted birds' bearings should have been deflected from those of the controls, but they were not. (4) Whatever cues were being used by the birds were ones that did not require time compensation because the bearings of control birds and clock-shifted birds were equally accurate.

Because our early tests were performed only 21 miles from the loft, an obvious first possibility was that, in the absence of the sun, the birds relied on familiar landmarks. To test this, we gave a group of young birds a series of 1- to 4-mile training flights (as many as possible of them under overcast) from all directions, followed by single 30-mile flights from north and south and single 20-mile flights from east and west. During all these training flights, and even during all exercise releases at the loft itself, a record was kept of how long each bird had been out of our sight at any one time. We then calculated from these data how close to the intended test site (102 miles east of the loft) each bird might conceivably have gotten if, while it was out of our sight, it had flown in a straight line at 60 mph toward the site and then turned and flown in a straight line 60 mph back to the loft. No bird that might have been closer than 50 miles from the test site under these unrealistically strict assumptions was used in the final tests. Yet the results of the tests at 102 miles were the same as those at 21 miles—under total overcast both controls and experimentals were oriented homeward and there was no difference between them. Therefore we concluded that pilotage by landmarks was not the back-up system used in the absence of the sun compass, but that the birds must be able to use alternative cues that permit them to navigate accurately when neither sun nor landmarks are available.

Since the original clock-shift tests under total overcast, we have performed more of the same sort from release sites in other directions. However, because of the great logistics problems of using the very severe criterion for unfamiliarity of release site described above, the criterion in these later tests has been that the birds have never previously been released within 35 miles of the test site. The results have been consistent: Under sun, the bearings of the experimental birds are deflected from those of the controls as predicted (Fig. 10A,B); under total overcast at "familiar" sites (i.e., ones where the 35-mile criterion is not met), the bearings of controls and experimentals are both homeward oriented (Fig. 10C,D); and under total overcast at unfamiliar sites, the bearings of controls and experimentals are still both homeward oriented (Fig. 10E,F).

Other investigators (Baldaccini et al., 1971; Walcott, personal communication) who have followed our training procedures have since also obtained good orientation under overcast. Wagner (1972), too, has reported "well-oriented vanishing diagrams even under complete overcast."

These results are, of course, completely at odds with the expectations of the sun navigation hypotheses of Matthews and Pennycuick. But, in addition, they raise serious questions about Kramer's map-and-compass model:

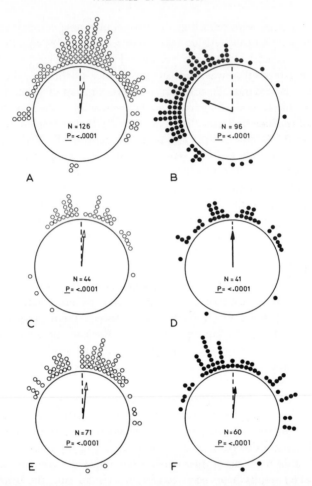

FIG. 10. Pooled vanishing bearings from releases of 6-hours-fast clock-shifted pigeons under sun (A,B); under total overcast at sites they might have seen before (C,D); and under total overcast at unfamiliar sites (E,F). The means under sun of controls (A) and experimentals (B) differ by 76°, whereas the corresponding differences in the other two cases are only 5° and 1.5°. (From Keeton, unpublished data.)

If the sun compass is not an essential element in the navigation process, does this mean that the "map" is capable in itself of providing sufficient information for true navigation, and does the sun compass simply function as an accessory check mechanism? If so, is the so-called map possibly not a single entity but a complex of cues whose grouping under a single term is perhaps misleading? Or can we retain the map-and-compass model intact by assuming that the birds have alternative compasses? Such questions as these must receive careful attention in the future.

III. RECENT DISCOVERIES CONCERNING POSSIBLE ORIENTATIONAL CUES

Both the finding that pigeons can navigate without benefit of the sun, even though they clearly use the sun when it is available, and numerous recent reports that migratory birds tracked by radar have been found to be flying in appropriate directions under dense cloud cover that would preclude use of the stars (e.g., Bellrose, 1967b; Drury and Nisbet, 1964; Nisbet and Drury, 1967; Steidinger, 1968; Gauthreaux, 1971; Evans, 1972; Richardson, 1971, and unpublished data), or even within cloud layers where neither the stars above nor the gound below would be visible (e.g., Bellrose and Graber, 1963; Williams *et al.*, 1972; Griffin, 1972, 1973), strongly suggest that birds can use a variety of alternative orientational cues. In short, like well-designed man-made devices, they possess back-up systems and alternative ways of weighting information inputs so that they can continue to function even when one or more of the preferred cues is unavailable. Furthermore, the cues utilized and the weighting scheme employed may well vary not only with environmental circumstances such as weather but also with the season of the year (Emlen, unpublished data), the age or experience of the individual (Keeton and Gobert, 1970; Keeton, 1972), the species of bird (Bellrose, 1972; Emlen, unpublished data), etc. The old idea that we will eventually discover *the* bird navigation system is thus giving way to a realization that we may discover a variety of interrelated systems.

Now, once we accept the idea that birds may be able to use alternative cues, depending on the circumstances involved, then we are forced to the conclusion that the results of many of the older investigations of possible use of various orientational cues may not be as conclusive as once thought. For example, if an experiment consisted in altering possible-cue *A* while keeping other conditions optimum, and if the birds continued to orient well, they may simply have used cue *B* as an alternative to *A*. Similarly, if we interfered with *B* while keeping all else optimum and the birds oriented well, they may have used *A* instead of *B*. We would, obviously, be entirely wrong if we interpreted these experiments as indicating that cues *A* and *B* are not elements in the birds' navigational system. In fact, the experiments have shown only that neither *A* nor *B* alone is *essential* for proper orientation under the particular test conditions.

It was reasoning of this sort, after our clock-shift tests had convinced us our pigeons could use alternative orientational procedures, that led us to begin conducting experiments in which we varied several possible cues simultaneously. We assumed that there must be limits to the redundancy of orientational cues, and that if we could interfere with enough cues at once we could hope to learn which are important and how they are related one to another. At about the same time, other researchers began looking more carefully at possible cues that had long been ignored or examined only superficially. In the sections below, I shall try to summarize the results of this recent work.

A. Magnetism

We have already seen (Section II,B,1) that the possibility that birds might obtain directional information from the earth's magnetic field has been around for a long time, but that after Yeagley's hypothesis was rejected the whole idea of use of magnetic cues in orientation fell into disrepute. There were several reasons for this: (1) Experiments in which magnets were placed on pigeons failed to cause disorientation (Casamajor, 1926; Gordon, 1948; Matthews, 1951b; van Riper and Kalmbach, 1952; Yeagley, 1951), except in a single release reported by Yeagley (1947). Similar tests on other bird species also yielded negative results (e.g., Wodzicki *et al.*, 1939; Matthews, 1952; Bochenski *et al.*, 1960). (2) Attempts to detect a sensitivity of birds to magnetic stimuli or to condition them to respond behaviorally to such stimuli were unsuccessful (e.g., Griffin, 1940, 1952a; Kramer, 1949, 1950a; Clark *et al.*, 1948; Orgel and Smith, 1954; Neville, 1955; Fromme, 1961; Meyer and Lambe, 1966; Emlen, 1970a). (3) Because the earth's magnetic field constitutes an environmental influence of which human beings are not normally conscious, it was easy to dismiss it as "having no known sensory properties," and as being "much too weak to be detected by a biological system."

But there were several indications in the literature that magnetism might sometimes constitute a biological stimulus after all. Frank Brown and his colleagues had published a series of papers reporting responses to magnetism by various invertebrates, such as protozoans, planarian worms, and mudsnails (summarized in Brown, 1971), though the effects were often so slight that there was no widespread acceptance of Brown's conclusions. There were also reports of responses to magnetism in a variety of insects (e.g., Becker, 1963a,b; Picton, 1966; Schneider, 1963; Lindauer and Martin, 1968) and in earthworms (Bennett and Huguenin, 1969). Furthermore, F. W. Merkel and his colleagues at Frankfurt (Merkel and Fromme, 1958; Merkel *et al.*, 1964; Merkel and Wiltschko, 1965; Wiltschko and Merkel, 1966; Wiltschko, 1968) had consistently reported that European robins in circular cages can maintain proper migratory orientation in the absence of visual cues, and that alteration of the directional component of the magnetic field surrounding their cages produces predictable changes in their orientation. Southern (1969a) had reported a correlation between the accuracy of the orientation of ring-billed gull chicks and fluctuations in the strength of the earth's magnetic field, as measured by the National Geomagnetic Observatory's K indices. And several Russian workers (Eldarov and Kholodov, 1964; Shumakov, 1965, 1967) had noted a general increase in locomotor activity when birds are put in magnetic fields of higher than normal intensity.

In view of this evidence that geomagnetism may have biological effects, we decided that magnetism would be the first possible cue investigated in our proposed "redundancy" experiments. First, we attached magnets to the backs of experimental birds and brasses to the backs of controls, and released the birds under sunny conditions at distances of 17–31 miles. The results agreed with those in the literature (cited above); both the experimentals and the controls were well oriented homeward and there was no significant difference between them (Fig. 11A,B) (Keeton, 1971a, 1972). But when experimental birds carrying magnets and control birds carrying brasses were released under total overcast, the results were quite different; the controls were usually still homeward oriented but the experimentals usually gave random bearings (Fig. 11C,D). This raised the possibility that the birds can use sun cues and magnetic cues interchangeably, but that both together are seldom needed.

It soon became apparent, however, that this possibility applies only to birds with some prior homing experience. First-flight youngsters (i.e., 3-months-old birds being taken away from home for the first time) were found to be unable to orient under overcast no matter how many exercise releases they had had at the loft (Keeton and Gobert, 1970), and furthermore they vanished randomly even under sunny conditions if magnets were attached to them (Fig. 11E,F) (Keeton, 1971a, 1972). Thus it appeared that very young inexperienced birds require both sun cues and magnetic cues to orient homeward. Perhaps one effect of early training is to make the birds more adept at homing so that they can orient with less information.

The disorienting effect of bar magnets on experienced birds released under overcast has now been confirmed by Papi et al. (personal communication). Southern (1972b) has found a similar effect on ring-billed gull fledglings under overcast.

Walcott (personal communication) has found that bar magnets sometimes have a disorienting effect on his experienced pigeons even when the sun is shining, if the release is at a long distance (138 miles). We, too, have found a very slight effect of magnets on experienced pigeons released under sun at long distances (Keeton, 1972). Walcott (1972a) has taken the matter a step farther by utilizing two small Helmholtz coils, one glued to the top of the bird's head and the other around its neck. When current from a battery pack on the pigeon's back is flowing through the coils, a magnetic field of approximately 0.1 gauss is produced in the region between the coils,

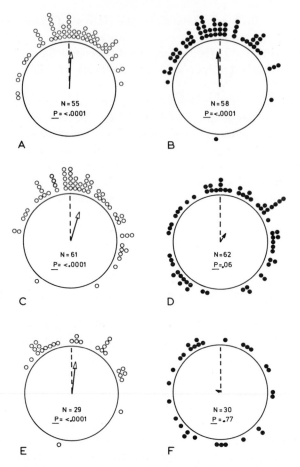

Fig. 11. Pooled bearings from tests comparing pigeons carrying brass bars (left) and magnet bars (right). (A,B) Under sun at 17–31 miles. (C,D) Under total overcast at 17–31 miles. (E,F) First-flight youngsters under sun at 17 miles. (From Keeton, 1972.)

where the bird's head is located.[5] Walcott has found that pigeons wearing this device with current turned on exhibit vanishing bearings under sun that are usually more scattered than the bearings of control birds wearing the same device but with no current flowing (Fig. 12).[6]

At the same time that these tests with magnets mounted on free-flying

[5] The values of 0.8–1.2 gauss given in Walcott's (1972a, p. 289) paper are misprints for 0.08–0.12 gauss.

[6] Recent dramatic results have come from experiments in which Walcott and Green (1974) compared the initial orientation of a group of pigeons wearing active coils with that of a second group also wearing active coils but with the direction of current flow reversed. The two configurations induced magnetic fields of the

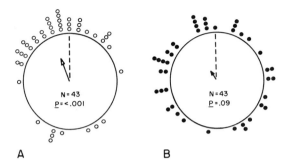

Fig. 12. Ten-mile bearings (radio) under sun of control (A) and experimental (B) pigeons wearing Helmholtz coils. The bearings of the experimentals were significantly more scattered. (From Walcott, 1972a).

birds were being conducted, Southern (1971, 1972a) was gathering more data on the orientation of ring-billed gull chicks in circular arenas. These new data greatly strengthened his earlier (Southern, 1969a) contention that the accuracy of the chicks' southeastward orientation declines progressively as the intensity of magnetic disturbance increases (K values 0–7).[7] Also Wiltschko and his colleagues were obtaining more data concerning the effects of magnetic fields of various strengths on the orientation of European robins under a variety of meteorological conditions (Wiltschko *et al.,* 1971;

same strength (approximately 0.6 gauss), but in the one case the field pointed up through the bird's head and in the other down. When released under sun, both groups oriented homeward, but under total overcast at unfamiliar release sites one group oriented homeward and the other oriented nearly straight away from home. These results support my conclusion that the importance of magnetic cues for experienced pigeons is greatest when sun compass information is unavailable. They further suggest that the field induced by the coils, which is far more uniform than that produced by the bar magnets I used, is capable of providing meaningful directional information to the birds, the two experimental configurations providing opposite information.

[7] The U.S. Coast and Geodetic Survey assigns a K value, ranging from 0 to 9, to each 3-hour period during the day. The K value indicates the intensity of magnetic disturbance, measured in gammas (1 gamma = 10^{-5} gauss). The lower limits, in gammas, for the K values (0–5,7) studied by Southern are 0, 5, 10, 20, 40, 70, 200. K values of 0–3 indicate very low levels of magnetic disturbance, 4 indicates weak magnetic storm activity, 5 indicates a moderate magnetic storm, and 6–7 indicate moderately severe storms; none of Southern's experiments took place under K values of 8 or 9, which would indicate severe storm activity.

Southern found strong southeastward orientation at K values of 0, 1, or 2, weak orientation at 3, and random bearings at 4, 5, and 7, with some tendency for reverse orientation (i.e., northward) at the latter two values. Since orientation was present at $K = 3$ and not at $K = 4$, and since these values differ in their lower limits by only 20 gamma, Southern's results suggest that gulls are sensitive to very small changes in the magnetic field.

Wiltschko and Höck, 1972). He also performed a series of experiments with different alterations of the polarity of the vertical and horizontal components of the magnetic field that led him (Wiltschko, 1972; Wiltschko and Wiltschko, 1972) to propose a mechanism whereby birds might tell compass directions by detecting the angle between the vectors of the earth's magnetic and gravitational fields.

It is important to say that our work on pigeons and Southern's on gulls has not definitely proved that these birds use magnetic cues in orientation, since the effects studied are primarily disorientation. However, it seems to me unlikely that the disorientation is due simply to some general physiological disturbance, since the effect in pigeons differs depending on weather conditions, the previous training of the birds, the age of the birds, and whether or not they have been at the release site before (Keeton, 1972). I think it more likely that the magnets have a direct effect on the orientational mechanism, and hence that pigeons are capable of using magnetic cues as one component in their orientation system. Certainly Merkel's and Wiltschko's tests with European robins appear to indicate a direct orientational response to the magnetic field.[8]

In principle, information from the earth's magnetic field could be used in a variety of ways by orienting birds. For example, the magnetic field is geographically quite irregular, with an abundance of major and minor anomalies, hence birds might become familiar with the magnetic topography, and use this in piloting. Or the magnetic field might provide one element in a hybrid bicoordinate system; e.g., the magnetic dip angle (inclination) might provide a rough estimate of latitude, which could be coupled with longitudinal information from some other source. Or, a third possibility, since the various parameters of the magnetic field (vertical intensity, horizontal intensity, dip angle, declination) do not vary completely in concert, two such parameters could serve as the elements of a bicoordinate navigational grid (though a very poor one since the divergence of the elements is not great). Or, finally, birds might use the magnetic field only as a simple compass. There is at present insufficient evidence to choose among these possibilities, though Wiltschko (1972) and Wiltschko and Wiltschko (1972) have argued that migrating European robins get compass information from the relationship between the vector of magnetic total intensity (without regard to its polarity) and the gravity vector. Wallraff (1972c) appears to have concluded that the magnetic field serves as a compass, since he says

[8] The results of Walcott and Green (1974) with pigeons also seem to indicate a directional response to magnetism. Furthermore, they appear to be consistent with Wiltschko's (1972) model in which north for a bird is the direction in which the magnetic vector forms the most acute angle with the gravity vector (i.e. with the vertical).

that, for direction orientation in birds, "three mechanisms are known . . . namely, the sun compass, the star compass, and the magnetic compass."

If birds really can obtain information from the earth's magnetic field, an important question to answer is how magnetic cues might be detected. Unfortunately, no answer is yet available. Some suggestions (e.g., Yeagley, 1947; Talkington, 1964) have been based on an induced emf in the bird as a consequence of its rapid flight through the earth's field. But Southern's gull chicks and Merkel's and Wiltschko's robins were not flying, and it seems unlikely that their movements in the circular arenas and cages were sufficiently fast to result in a significant induced emf. I therefore lean to the idea that the detection is based on some sort of direct effect of the magnetic field on a sensory apparatus.

But what might that sensory apparatus be? And what is its most likely location? We are accustomed to thinking of detectors for external stimuli as usually being on or near the surface of the body, but this need not be the case with magnetic detectors, because magnetic lines of force can pass right through tissues. Hence the detectors could be anywhere inside the animal, such as in the circulatory system or even in the brain itself.

An important point to keep in mind while searching for magnetic detectors is that the direct effect of the magnetic stimulus may be very small and thus easily overlooked. Just as movement of the cilium of a hair cell in the cochlea through a distance less than the diameter of a hydrogen atom is sufficient to initiate impulses in the auditory nerve fiber, so a tiny change in the distribution of ions in a membrane, or a minute Hall-effect voltage resulting from interaction of the magnetic field and the DC potential of a tissue or nerve cell (which may have some semiconducting properties), might be quite enough to result in informational input for the bird. Or, to cite another example, it has been shown recently (Chalazonitis *et al.*, 1970) that a strong magnetic field causes rotation and alignment of suspended retinal rod cells, and Hong *et al.* (1971) have suggested this is due to the oriented anisotropy of the regularly arranged membranous lamellae in such cells. Now, the earth's magnetic field, being only approximately half a gauss, would surely not produce much change in alignment in the disc-lamellae of intact rod cells; nonetheless a reorientation so small as to have escaped attention might be all it would take to trigger a sensory input. My point here is not to espouse an ionic redistribution or a Hall effect or a retinal-cell change as the magnetic detection system, but simply to call attention to the fact that we should not necessarily seek some gross effect of magnetism on the organism as the mechanism of detection.

Recent research has provided a few intriguing hints that may prove helpful in our search for magnetic detectors. I shall mention two here. First, it appears that the system is slow to respond to any major change in the magnetic stimulus. Thus Wiltschko (1972) has found that his robins, accustomed to the normal Frankfurt field of 0.46 gauss, initially give random bearings when exposed to fields 26% lower (0.35 gauss) or 48% higher (0.68 gauss), but the birds will orient in a field as low as 0.16 gauss or as high as 0.81 gauss if they are first kept in the altered field for 3 days. A similar sensitivity to altered field strengths has been found in whitethroats (*Sylvia communis*) (Wiltschko and Merkel, 1971). In their elegant studies of orientational responses of honeybees to the earth's magnetic field, Lindauer and Martin (1968, 1972) have found that when they artificially compensate the magnetic field (to 0.4% of its normal strength), it takes 30–45 minutes before the bees begin performing "error free" dances. They have also found that the bees show a pronounced lag in responding to rapid fluctuations in the magnetic field (more than 1 gamma per 1° of sun azimuth change), whether naturally or artificially produced.[9] Brown (1971), discussing the effect on mud snails of an experimental magnetic field deviating in strength from the earth's, reports that the effect "persisted for at least five to ten minutes after the experimental field was removed." And in studies of the effects on planarian orientation of magnetic fields stronger or weaker than the normal one, he reports that persistent effects were evident for 25–30 minutes after the artificial fields were removed. I think we can reasonably conclude from these data that the magnetic detectors must be slow adapting, and this fact should help us evaluate the likelihood of various possible mechanisms that may be proposed.

The slowness of the response to an altered magnetic field may be the answer to why so many investigators have failed in their attempts to condition birds to magnetic stimuli. The stimuli have often been quick pulses or flashes (e.g., Emlen, 1970a), or else the animal has been put in a maze or key-pecking apparatus where it is exposed to the altered field for a relatively brief period. Only Reille (1968) has reported positive results (autonomic conditioning of an increased heart-rate response to an abrupt, but short, change in the direction of the horizontal component of the magnetic field used as a warning that the bird is about to be shocked), but we (Kreithen and Keeton, 1974c) have repeated Reille's experiments with negative results. We are now attempting to design conditioning tests that will take into account the possibility that the bird may need a half hour or more of exposure to an altered field before responding.

A second hint (though it is perhaps not as convincing as the previous

[9] Lindauer and Martin (1972) report data indicating that bees are sensitive to changes in the magnetic field of only 1–300 gamma. These investigators are now convinced bees can detect magnetic changes of less than 5 gamma (personal communication).

one) is that magnetic detection may be linked in some way with gravity detection. Thus Lindauer and Martin (1968, 1972) have shown that the *Missweisung* of honeybees on a vertical comb is zero when the magnetic vector coincides with the appropriate dance vector, which is oriented relative to gravity, and at its maximum when the magnetic and gravity-dance vectors are most divergent. A similar effect of the magnetic field on a response to gravity has recently been demonstrated in *Drosophila* (Wehner and Labhart, 1970). Wiltschko and Wiltschko (1972) have proposed that European robins can tell compass directions from the position of the angle between the magnetic and gravity vectors.

B. INERTIAL GUIDANCE

In this day of sophisticated rocketry and space probes, the possibility that birds might use some sort of inertial guidance comes immediately to mind. Indeed, the possibility that birds might detect all the twists and turns of the outward journey and use these to calculate the route home was suggested as early as 1873 by Charles Darwin.

Homing by inertial information could conceivably take either of two forms. A bird might simply retrace his outward journey, or he might double integrate all the acceleration data and thus compute the direct route home. The first of these possibilities can be ruled out immediately on the basis of hundreds of experiments that have shown that: (1) pigeons carried to release sites by quite lengthy and circuitous routes return home much faster than would be possible if they had to retrace the outward journey; (2) when pigeons are brought to a release site by a variety of routes that approach the site from different directions, their initial departure bearings show no correlation with the approach bearings;[10] (3) when the actual homing flights are mapped by airplane tracking, the routes bear no relationship to the routes followed on the outward journey.

The second possibility has been harder to rule out to everyone's satisfaction. In fact, Barlow (1964, 1966) has been a strong proponent of the idea that birds might navigate in this way, and he (Barlow, 1971) has recently stated that "the really relevant experiments perhaps have yet to be carried out." However, it is certainly true that the available experimental evidence is overwhelmingly negative. This evidence is of four sorts:

(1) Birds have been carried to release sites while being rotated on turntables (Rüppell, 1936; Griffin, 1940; K. E. Money, personal communication; Covey and Keeton, unpublished results) or in lightproof drums (Mat-

---

[10] For a recent report that the direction of the first part of the outward journey influences initial orientation at the release point, see Papi *et al.* (1973b). However, my attempts to repeat this work consistently yielded negative results (Keeton, unpublished).

thews, 1951b). The assumption in such experiments is, of course, that the hypothesized detection and integration systems cannot be 100% accurate, and that if there is enough noise in the input, the cumulative error must eventually become large enough to destroy any possibility of locating the homeward direction by inertial means. However, no difference between experimental and control birds in either initial bearings or homing success has been found in such experiments.

(2) Birds have been carried to release sites while under deep anesthesia (Exner, 1893; Kluijver, 1935; Griffin, 1943; Walcott and Schmidt-Koenig, 1973). When released after recovery, the experimental birds have not differed significantly from the controls in either initial bearings or homing success.

(3) Birds with surgical lesions of the vestibular organs—the presumed principal detectors of accelerations—have been tested (Hachet-Souplet, 1911, horizontal semicircular canals bisected; Sobol, 1930, horizontal canals bisected; Huizinger, 1935, all canals cut in some birds, only the horizontal canals cut in others, the pars inferior [sacculus, cochlea, and lagena] and the ampullae of the posterior canals removed in others; Wallraff, 1965, horizontal canals bisected; Wallraff, 1972a, cochlea and lagenae removed; Money, personal communication, sacculi removed in some birds, lagenae removed in others). The results have been consistently negative, i.e., there have been no significant differences between experimental and control birds (Fig. 13A). Since all these tests were presumably performed under sunny conditions, and since our current appreciation of the redundancy in the pigeon navigation system has made us aware of the need for varying several possible cues simultaneously, K. E. Money and I (unpublished results) have recently conducted tests under total overcast at unfamiliar release sites, using pigeons whose sacculi have been removed. Again the results were negative—

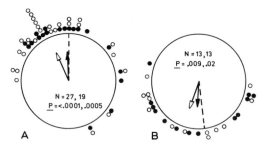

FIG. 13. Bearings of pigeons with surgical lesions of the vestibular apparatus. (A) Pooled bearings from Wallraff's (1965) tests of birds with the horizontal semicircular canals bisected. (B) Bearings from a test under total overcast of birds with the sacculi removed, 47 miles north of Ithaca, N.Y. (Money and Keeton, unpublished data.) Both the controls and the experimentals were significantly homeward-oriented in each case.

the experimental birds performed just as well as controls in both initial bearings (Fig. 13B) and homing success.

(4) The measured sensitivities of the vestibular system to angular and rectilinear accelerations, respectively, are approximately $0.2°/sec^2$ and 6 cm/sec^2 (Barlow, 1964), which is not sufficient for accurate localization of home after displacement.

Taken as a whole, the above evidence has convinced most students of avian orientation that calculation of the homeward direction from inertial information obtained on the outward journey is not likely. But as long as the avian navigation system (or systems) remains as poorly understood as it is now, it behooves us at least to keep the possibility, no matter how unlikely, in mind. We need only recall the too ready rejection of a role for magnetic cues a few years ago. Certainly a confirmed proponent of inertial guidance can still claim that: (1) the birds' detection and integration systems are far more accurate than measurements have indicated—so accurate, in fact, that they can adequately cope with the track distortions created by a ride on a turntable; (2) anesthesia does not make all the nervous system cease functioning, hence perhaps the parts concerned with inertial detection continue operating in an unconscious bird: (3) perhaps the labyrinths are not the chief sensors of acceleration information for inertial guidance purposes. In short, we are up against an old problem—negative evidence, no matter how voluminous, cannot absolutely rule out the possibility that a given source of information might be used in some circumstances. It's impossible to prove that something is impossible.

Having revealed my own bias against the likelihood of use of inertial information obtained during the outward journey, let me hasten to say that I think we should look carefully, in future experiments, at the possibility that some birds may use inertial information as an aid in maintaining a straight course once they have chosen a bearing and are flying along it (Drury and Nisbet, 1964). This simpler use of inertial information makes less demand for accuracy on the vestibular system and is therefore more in line with the measured sensitivities (though the lesion experiments mentioned above cast doubt on even this possibility in pigeons). The findings of Schreiber et al. (1962) that racing pigeons and migratory doves show prolonged cerebellar after-discharges following rotation on a turntable, whereas ordinary wild pigeons and nonmigratory doves do not, are suggestive in this regard, but they may merely be an indication of the special demands imposed on the cerebellum by the aerodynamic problems of extended flight.

C. Stars

The use of stellar cues as a source of orientational information has been well documented in a variety of nocturnal migrants. Since the Sauers'

(Sauer, 1957, 1961; Sauer and Sauer, 1959, 1960) studies on European warblers, the list of species has been greatly extended, and now includes ducks (Bellrose, 1958, 1963; Hamilton, 1962a; Matthews, 1961, 1963a), sandpipers (Sauer, 1963; Emlen, personal communication), and many different songbirds (e.g., Emlen, 1967a,c; Hamilton, 1962b, 1966; Mewaldt and Rose, 1960; Mewaldt et al., 1964; Rabøl, 1969, 1970; Shumakov, 1965, 1967).

Sauer's work suggested that some of his warblers might be selecting a particular star or group of stars and orienting by flying at an appropriate angle to it. This would require the integration of an internal time sense, because the angle of orientation relative to the selected star would have to be compensated to allow for the apparent rotation of the sky. By contrast, Emlen (1967b) clearly demonstrated that his indigo buntings use configurational information, i.e., they use patterns of stars to locate a directional reference point, a process that does not require time compensation. Emlen (1969a,b, 1970b, 1972) has examined in considerable detail the ontogenetic development and seasonal timing of the stellar orientation mechanism in buntings.

In my opinion, Emlen's experiments strongly suggest that buntings use the stars as a compass only, not as a basis for bicoordinate navigation. Sauer's experiments on European warblers may point to a bicoordinate system (Sauer and Sauer, 1960; Sauer, 1961) but the data are much too few to be convincing (Wallraff, 1960b,c; Emlen, in preparation). Neither Matthews (1963a) nor Wallraff (1969), working with stellar orientation by mallards, could find any evidence of time compensation. Though it is entirely possible that different species of birds may employ very different celestial systems, I must conclude that thus far only the compass system has been established. I conclude also that use of stellar information in homing, as distinguished from maintenance of a migratory course, has not been demonstrated, although it would seem worth investigating whether such a process might come into play near the end of a migratory journey, when localization of a specific breeding or wintering spot occurs.

The subject of avian stellar orientation is much too big to be covered adequately in the present paper, since its direct relevance to homing is not clear. For a full and careful review of this important topic, the reader is referred to Emlen (1975).

D. WALLRAFF'S "ATMOSPHERIC FACTOR"

As early as 1954, Kramer and von St. Paul conducted homing tests with aviary pigeons, i.e., birds that had lived in wire aviaries since leaving the nest and had never been released for exercise at the home location (Kramer and von St. Paul, 1954). They found that, when released at distant sites, such pigeons showed a homeward tendency, as indicated by both the initial bearings and the distribution of locations where they were later found (e.g., see Fig. 2, which shows results of a test of this type conducted by Wallraff). Shortly before he died, Kramer (1959; Kramer *et al.,* 1959) began a series of tests in which he varied what the pigeons could see from their aviary. When the aviary was surrounded by wooden "palisade" walls that were a little higher than the aviary itself (the birds' view from the highest perches was blocked to a point 3° above the horizontal), all indication of homeward orientation disappeared.

Wallraff (1966b, 1970a) has continued the studies of aviary pigeons, taking up the subject where Kramer left off. His results are interesting but, to me at least, extremely confusing. I shall not attempt here to cover all the different types of experiments Wallraff has performed, but shall mention those I consider most important.

After confirming that his wire-aviary pigeons, like Kramer's, oriented toward home upon release at a distant site (e.g., Fig. 14A), and that pigeons whose view of the surroundings was blocked by wooden walls up to a point a few degrees above the horizon did not, Wallraff next tried a so-called roofed palisade. Wooden walls surrounded the aviary and a wooden roof

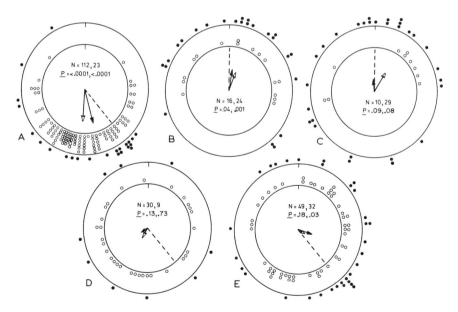

FIG. 14. Initial bearings (inner circles) and recoveries (outer circles) for pigeons from Wallraff's wire aviary (A); roofed palisade (B); glass palisade, 1966 (C); glass palisade, 1970 (D); louver palisade (E). (After Wallraff, 1966b, 1970a.)

covered it, but between the top of the walls and the edge of the roof was a narrow slit through which birds on the highest perches could see the horizon and the lowest parts of the sky. The slit was closed during early morning and late afternoon so that the birds never had a direct view of the sun. When released at a distant site, pigeons from this roofed palisade showed homeward orientation in both initial bearings and recoveries (Fig. 14B). Since the birds in the roofed palisade appeared to have gotten, from the horizon, information that the birds in the ordinary wooden palisade had not gotten from the entire sky except the 3° at the horizon, Wallraff (1966b) concluded that there is some sort of "substrate of information spreading in the horizontal plane which is important for homing ability."

That this horizon-factor information must not be visually perceived was indicated by experiments in which Walraff (1966b, 1970a) substituted glass walls for the upper part of the wooden ones of the original palisade, and found that the orientation of the birds (Figs. 14C,D) was poor. In later experiments, Wallraff (1970a) tested birds from a square aviary that was enclosed on its north and south sides by wooden walls and on its east and west sides by walls composed of plastic louvers that blocked vision but permitted air flow. He reported that these birds could orient when tested (Fig. 14E), thus indicating that nonvisual "oriented dynamic processes in the atmosphere are involved in the navigational system of homing pigeons." Thinking that the atmospheric factor might be infrasound, "dynamic structures of air currents," or some other parameter detected by the bird's ear, Wallraff (1972a) extirpated the cochleae and lagenae, but found that this caused no detectable decrement in homing ability.

I have already confessed that I find Wallraff's findings confusing. I must say, further, that in some instances I am not sure I find the evidence quite as convincing as he does. There seems no doubt that wire-aviary pigeons are capable of homeward orientation, and Wallraff (1970b) has shown that this technique can be used profitably in studying other aspects of homing, such as the special properties of individual release sites. Moreover, I am reasonably convinced that roofed-palisade pigeons are also capable of homeward orientation. But I have some difficulty when comparing the performances of glass-palisade and louver-palisade birds. Thus, when I reanalyze (Fig. 14C, inner circle) the glass-palisade data in Wallraff's (1966b) paper, utilizing only the "good" bearings, i.e., ones where the birds could be followed to the limits of vision,[11] I find that, though they are statistically ran-

---

[11] Many of the palisade birds fly very low and soon land or are lost from view behind trees, low hills, or other obstructions before they have truly departed. Such bearings would seem to be meaningless and therefore properly omitted from our analysis. Unfortunately the meaningless bearings cannot be omitted from our analysis of the 1970 glass-palisade and louver-palisade data, because Wallraff does not provide the necessary information.

dom under the Rayleigh test ($P = 0.089$), they are nonrandom homeward ($P = 0.030$) under the V test; I can't help wondering if a larger sample size (the N was 10) might have yielded significance under the Rayleigh test also. Similarly, the bearings of the recoveries of these birds (Fig. 14C, outer circle), though statistically random ($P = 0.076$) under the Rayleigh test, are oriented nonrandomly homeward ($P = 0.012$) under the V test. The glass-palisade data in Wallraff's (1970a) paper (Fig. 14D) are more clearly random in both initial bearings and recoveries (Rayleigh $P = 0.129$ and 0.730, respectively).[12] But if we turn to the initial bearings of the louver-palisade birds (Fig. 14E, inner circle), we find them random under the Rayleigh test at $P = 0.18$, which is a poorer value than we obtained for the glass-palisade birds. Thus the argument that the louver-palisade birds received some sort of nonvisual, horizontally spreading, dynamic atmospheric information that was blocked by glass rests almost entirely on the fact that the recoveries of the louver-palisade birds (Fig. 14E, outer circle) were oriented nonrandomly homeward (Rayleigh $P = 0.027$, V test $P = 0.011$). Since we have seen that in one of the two sets of data on glass-palisade birds the recoveries were oriented homeward under the V test and nearly so under the Rayleigh test, and since recoveries are, at best, only a rough indication of orientation because they depend so much on chance events, such as the geographic distribution of pigeon fanciers and whether or not a lost pigeon enters a loft rather than joining a wild flock, I feel Wallraff's assertion that the glass-palisade and louver-palisade birds behaved differently remains to be proved.

E. OLFACTORY CUES

It has been traditional to regard most birds as having a very poor sense of smell, but recent research (e.g., Tucker, 1965; Henton, 1966; Henton et al., 1966; Shumake et al., 1969; summary in Wenzel, 1971a) has indicated that most birds have some olfactory sensitivity and that some birds have a keen sense of smell which they probably use in locating food or nests (e.g., turkey vultures, Stager, 1964; the kiwi, Wenzel, 1971b; petrels and shearwaters, Grubb, 1972). The size of the pigeon olfactory bulb, relative to the cerebral hemispheres, is intermediate between that of birds that probably have a very good olfactory sense (e.g., kiwi, Procellariiformes, Podicipediformes) and birds that probably have a poor olfactory sense [e.g., most Passeriformes (Bang and Cobb, 1968)], hence the possibility that olfactory information may be important to pigeons in some contexts cannot be ruled out.

[12] Though the final vanishing bearings of these glass-palisade birds are random at $P = 0.129$, the bearings after 20 seconds and after 40 seconds are nonrandom, whereas the 20-second and 40-second bearings of the louver-palisade birds are random.

Papi *et al.* (1971a, 1972) have recently reported that pigeons whose olfactory nerves have been sectioned, or whose nostrils have been plugged with cotton, give random vanishing bearings at release sites and have very poor homing success. On the basis of these results, Papi *et al.* (1972) have proposed a new hypothesis of pigeon homing, which they feel takes account of all that is known to date. The essential points of their hypothesis are as follows: (1) Volatile odorant substances, presumably organic, give rise in each area to a distinctive pattern of olfactory stimulation. (2) Young pigeons learn the odor of the loft area, and, when winds blow from different directions, they also learn to associate different odors with different directions as determined by their sun or magnetic compasses. (3) When released at an unfamiliar site, the birds detect the odor of that site and, if it is one they have previously experienced at the loft, they establish the home direction as opposite to the one from which the odor had usually come to the loft. (4) The birds then locate the deduced direction by means of their sun or magnetic compass, and fly home. (5) Experience improves the birds' homing ability by making them more familiar with the odor patterns characterizing various areas they encounter, and perhaps by familiarizing them with landmarks they can use as auxiliary cues. Papi *et al.* (1972) feel that their hypothesis is consistent with Kramer's map-and-compass model in that it envisions a two-stage orientation process, but their map component gives only home direction, not distance. They also suggest that their hypothesis would explain the results of Wallraff's various palisade experiments by identifying odorous substances as the atmospheric-borne factor he postulated.

It is too early to evaluate satisfactorily all aspects of this hypothesis. However, a few points may be worth mentioning: (1) It is difficult to imagine that homing from sites far beyond any previously visited could be explained in this way, as Papi *et al.* themselves admit. Since inexperienced pigeons can orient homeward at very distant release sites (e.g., Wallraff, 1970b), we would thus have to assume that the birds use a nonolfactory navigation system under such circumstances. But if they possess such a system, why should they not use it at shorter distances? To me, there is no convincing evidence that pigeons use entirely different navigation systems at short and long distances. (2) Mixing of odorants by winds would seem to make accuracy in correlating odors with directions difficult, particularly in the case of odors from distant sites that may reach the loft by very circuitous paths. (3) If odor detection were really to be the basis for the remarkable homing ability of pigeons, one would expect anatomical evidence of a highly developed olfactory system, but the evidence actually indicates only a moderately developed system (Bang and Cobb, 1968). (4) When surgical intervention results in a behavioral deficit, it is always exceedingly difficult to determine whether the effect is specifically related to the model the experimenter had in mind (in this case, orientation by olfaction) or whether it is due to some more general disturbance. That Papi *et al.* saw no disturbance of reproductive and maintenance behavior in the operated birds may not be entirely relevant to the question of the possible effects of such an operation when heavy demands are being placed on the birds' respiratory system, as in flying. (5) The cotton plugs inserted in the nostrils of some of the experimental birds would be expected to impose unusual stress in view of the respiratory demands of flight; hence the atypical behavior of these birds does not seem to me to be convincing evidence of lack of orientational ability. (6) If future research should confirm a specific effect of olfactory nerve bisection on orientation, it would then be necessary to determine whether this is due to interference with olfactory input or to interference with detection of high-energy radiations. It has been shown that in some animals, including pigeons (Smith and Tucker, 1969), the olfactory mucosa is a major site of X-ray detection.[13]

F. METEOROLOGICAL CUES

There is a considerable body of evidence that many migrating birds possess the ability to detect and respond to favorable weather conditions. Thus the bulk of migratory movement in any particular area occurs on only a small percentage of the days or nights (Bagg *et al.,* 1950; Lack, 1960; Drury and Keith, 1962), and these times of dense migration are predictable from meteorological indicators (Nisbet and Drury, 1968; Richardson, 1971, 1972). This selectivity on the part of the birds is probably more than simply a way of avoiding such stressful circumstances as fog, heavy rain, and snowstorms. Since the metabolic costs of flight are very high, migrants that could correctly predict the winds aloft while still on the ground and initiate migratory flights only when those winds would be roughly following ones, could thereby effect very significant energetic savings. The birds' remarkable abilities as weather prophets may be due, in part, to their very great sensitivity in detecting changes in barometric pressure (Kreithen and Keeton, 1974a). Once aloft, the birds appear to seek out the favorable winds and, if the wind directions are different at different altitudes, to concentrate at the appropriate levels (Bellrose, 1967a; Blokpoel, 1970; Richardson, 1971; Steidinger, 1972; Bruderer and Steidinger, 1972).

Now, the idea that migrants actively seek winds blowing in roughly the correct direction implies that the birds must first determine the proper migratory direction, i.e., they must orient by other cues before selecting favorable winds. But if birds are so sensitive to winds, the possibility arises

[13] Papi *et al.* (1973a, 1973b) and Benvenuti *et al.* (1973a, 1973b) have recently published additional results from their investigation of the possible role of olfaction in pigeon homing.

that they might use them more directly as a source of directional information (Griffin, 1969). Indeed, Vleugel (1954, 1959, 1962) proposed that nocturnal migrants orient initially by observing the sunset direction, and then maintain the selected course during the night by flying at a constant angle to the wind. One might at first think that this would be possible only if the birds could see the ground, so that the direction and speed of the wind could be monitored visually. But, in fact, air flow at the altitudes where birds fly is characterized by patterns of turbulence that are related to the wind direction (Nisbet, 1955; Griffin, 1969). Bellrose (1967a,b) has argued that these patterns provide enough information to enable migrants flying under or in dense clouds to maintain their course.

The question whether selection of favorable winds after orientation by other means is sufficient to explain the marked downwind tendency of migratory flights (especially of passerines) or whether the wind itself is the dominant directional cue (as urged by Gauthreaux and Able, 1970; Able, 1974) has proved difficult to resolve. A relevant point is whether migrants compensate for wind drift, i.e., whether they try to maintain a constant flight direction by adjusting their heading to compensate for lateral displacement by the wind. This, too, is a controversial issue, but there is good evidence for compensation in at least some migrant species, especially strong flyers such as shore birds and waterfowl, and even small passerines in winds of low or moderate velocity (e.g., Evans, 1966, 1972; Drury and Nisbet, 1964; Nisbet and Drury, 1967; Bellrose, 1967a; Lack, 1969; Steidinger, 1972; Gauthreaux, 1972). For a more detailed discussion of the role of meteorological factors in migratory orientation, see Emlen (in preparation).

This brief summary has been included here because meteorological factors, particularly prevailing wind patterns, could potentially provide information for homeward orientation in some instances. For example, oceanic birds might conceivably use wind cues as one element in their navigational system. However, I know of no convincing evidence that wind direction *per se* greatly influences initial orientation of homing pigeons, except for a slightly increased scatter when the birds must fight a very strong side- or headwind. In tests of pigeons at the same distant unfamiliar release sites under quite different winds, the mean bearings are usually very similar (e.g., Wallraff, 1960a; Keeton, 1969).

G. FAMILIAR LANDMARKS

I have already (Section II,A) mentioned evidence that pilotage by familiar landmarks cannot explain orientation by inexperienced birds at distant release sites (e.g., Figs. 1, 2, 20B). But this does not necessarily mean that such landmarks might not be important aids when they are available. Downhower and Windsor (1971) have, in fact, recently published evidence suggesting that bank swallows may sometimes use landmarks located within 3 miles of their home colony. Indeed, many investigators have assumed that birds' navigation systems couldn't possibly be accurate enough to get them closer to home than 15 to 30 miles, and that this last portion of the homeward voyage must therefore be accomplished by pilotage. However, the evidence now suggests that landmarks, even when available, are given minimal weight in the hierarchy of orientational cues used by pigeons, and, moreover, that the assumption that homing from points near the loft must be by pilotage may be incorrect.

Schmidt-Koenig (1965, see for specific references), summarizing the evidence available at that time against an important role for landmarks, mentioned the following points: (1) At some release sites, initial bearings deviating significantly from the true home direction persist in very experienced pigeons even after repeated releases at the same sites. (2) Marked day-to-day changes in initial orientation may occur at sites where the birds have been released many times. (3) Differences in homing success have been observed in simultaneous releases of experienced pigeons from four directions at distances of only 3 miles from the loft. (4) Resetting of the birds' internal clock can result in deflected initial bearings at release sites as close as 1.24 miles (see Schmidt-Koenig, 1965, Fig. 16).

More recent evidence is of several types:

a. Airplane Tracking. Michener and Walcott (1966, 1967a) and Walcott and Michener (1967) have used an airplane to track pigeons carrying radio transmitters. They report, "We found no evidence to indicate that our pigeons piloted their courses by familiar landmarks." [But see a criticism of this work by Murray (1967) and the reply by Michener and Walcott (1967b).] In particular, they emphasize that, in 104 instances, they were tracking a pigeon flying an "incorrect" course when it encountered a path over with it had recently flown. The percentage of such birds that turned toward home upon encountering the "familiar" territory was no higher than would be expected if course changes occurred entirely at random. Michener and Walcott (1967a) did conclude, however, that their pigeons probably switched to landmark pilotage when within 5 to 10 miles of the loft.

b. Clock-Shift Tests at Special Training Sites. It has been shown (Keeton, 1969) that pigeons released daily for 2 weeks at the same site will, upon being clock-shifted 6 hours and released at that site under sun, choose bearings roughly 90° different from the controls and, in general, not get home the day of release. In other words, extensive previous experience at the site, with all the attendant opportunity for learning landmarks, does not prevent clock-shifting and the resulting erroneous interpretation of solar information from sending the birds in an incorrect direction.

   c.   Clock-Shift Tests Very Near the Loft. Graue (1963) [14] conducted test
releases of 6-hour clock-shifted pigeons at four release sites a mile or less
from the loft, i.e., in the area the birds flew over nearly every day in the
course of their natural activities. He found that at two of the sites (0.8
mile E, 1 mile SE), where the loft building was clearly in view, both experimental and control birds flew directly to the loft. By contrast, at the other
two sites (0.8 mile N, 0.6 mile WNW), where trees blocked a direct view
of the loft, many of the experimental birds flew off in the direction appropriate to their clock-shift (i.e., roughly 90° to the left or right of home)
whereas the control birds flew straight home. Graue concluded that a pigeon
"evidently uses the sight of the loft to guide its homeward flight, but it
does not use other adjacent landmarks, such as the woods, in the same way."

   I have already mentioned that Schmidt-Koenig (1965) got similar results
at 1.24 miles. More recently Schmidt-Koenig (1972, discussion) has said
that in clock-shift releases 2 km from the loft, "Although they could see
the loft very well many experimentals headed away from the loft." No data
have yet been published to support this statement, but if the report is correct,
we must ask why a direct view of the loft building itself took precedence
over other cues in Graue's experiments and not in Schmidt-Koenig's.

   Judith R. Alexander (1974), recently performed a series of clock-shift
releases at sites close to our loft at Ithaca, New York. A principal aim of
her work was to establish whether a direct view of the loft cancels the de-
flecting effect of clock shifts, as Graue's work suggests. She found that at
a release site 0.55 mile S of the loft, where the loft building is clearly visible,
most experimental birds flew straight home but a few were deflected in a
manner indicating an effect of the clock-shift (Fig. 15A). At a site 0.78
mile NW, where the loft is clearly visible at the altitude pigeons fly (though
hidden by a grove of trees from observers on the ground), the experimentals
vanished in the shifted direction in one test but not in another test (Figs.
15B,C); there was no obvious reason for the difference in results on the
two days. At a site 0.9 mile E, experimentals sometimes gave shifted bearings
and sometimes not; groves of trees stand between this release site and the
loft, and observers had the impression that it was the low-flying birds, which
probably could not see the loft building (though they certainly could see
a large laboratory building standing 110 yards from the loft), that vanished
in the shifted direction, and that it was the high-flying ones, which may
have been able to see the loft, that vanished directly toward home. At a
fourth site, 0.9 mile W, where only very high-flying pigeons would be able
to see the loft because of intervening trees, the experimentals consistently
vanished in the shifted direction (Fig. 15D). The overall result of this series

[14] Though Graue's paper was published in 1963, it was not mentioned in Schmidt-
Koenig's (1965) review, so it is treated in some detail here.

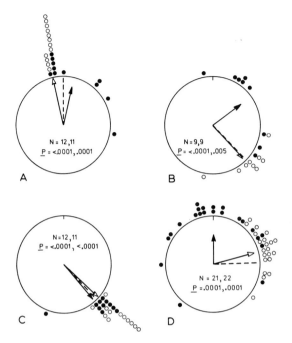

Fig. 15. Bearings of 6-hours-fast clock-shifted pigeons released less than a mile from the Cornell loft. (A) From 0.55 mile south. (B) From 0.78 mile northwest, August 11, 1970. (C) From 0.78 mile northwest, August 17, 1970. (D) From 0.9 mile west. The loft building was visible in the first three releases but not in the last. (From Alexander, 1974.)

of tests seemed to be a confirmation that, at release sites less than a mile from the loft, a direct view of the loft building itself usually overrides the effect of a clock shift, though a view of nearby buildings does not. But the one test in which the bearings of shifted birds were deflected (Fig. 15B) at the NW release site, together with the several deflected bearings at the S site (Fig. 15A), suggest that sometimes, for reasons not yet known, a direct view of the loft is ignored.

Figure 16 shows examples of the results Alexander has obtained in clock-shift tests at two release sites 3.7 miles NNW and 3.5 miles E. The loft building can be seen by the human eye from both sites, yet the experimental birds vanish in the shifted directions. However, these results cannot be in-terpreted as evidence that at this distance pigeons ignore a direct view of the loft, because if the latest evidence (Blough, 1971) indicating the visual acuity of homing pigeons to be considerably less than that of human beings (1.9–4.0 minutes of arc versus approximately 0.8 minutes of arc) is correct, then calculations show that the loft building is not large enough to be seen

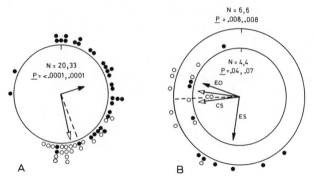

Fig. 16. Bearings of 6-hours-fast clock-shifted pigeons released 3.7 miles NNW (A) and 3.5 miles E (B), at sites where the loft building is visible to the human eye. The first part of the (B) release was conducted under sun (outer circle), but the last part was under total overcast (inner circle); note that under sun the controls, CS, were homeward-oriented and the experimentals, ES, were deflected southward, whereas under overcast both the controls, CO, and the experimentals, EO, headed homeward. (Adapted from Alexander, 1974.)

by the pigeons at distances of 3.7 and 3.5 miles. It remains to be determined, therefore, whether there are intermediate distances (e.g., 1 to 2 miles) where pigeons can see the loft but nonetheless consistently ignore it in choosing a departure direction. We are currently conducting experiments designed to answer this question.

Taken together, the findings of Graue, Schmidt-Koenig, and Alexander appear to establish conclusively that even in the immediate vicinity of their home loft pigeons do not ordinarily pilot by familiar landmarks. Graue (1963) has suggested that, even though the birds don't pilot by local land-marks, they do use them to determine their position before setting a course with the sun compass. However, he offers no evidence for this except his feeling that "since the distance involved was so very short, it would not seem reasonable to believe that the long distance orientation mechanism could have been involved. . . ." Though Matthews (1968, p. 85) has readily ac-cepted Graue's conclusion and supports the idea of "translation of landmark observation into compass directions," I find it entirely unconvincing; there is absolutely no evidence to support it, and I differ entirely with these authors as to what is "reasonable" and what is not in this case. All investiga-tors to date (e.g., see Schmidt-Koenig, 1972) have found that the behavior of clock-shifted pigeons released near the loft appears in no fundamental way different from that of clock-shifted pigeons released at very distant un-familiar sites, where Matthews himself would insist the birds orient by an entirely different mechanism. The same holds true for orientation under other experimental conditions, as summarized by Schmidt-Koenig (1965).

Therefore, to me the most "reasonable" conclusion is that the birds use their true navigation system until they are in sight of the loft, rather than switching to pilotage.

d. Experiments With Contact Lenses. Schlichte and Schmidt-Koenig (1971), Schmidt-Koenig and Schlichte (1972), and Schlichte (1973) have recently published the results of experimental releases they have conducted with pigeons wearing frosted contact lenses that admit some light but prevent image vision of objects more than a few meters away. When released at distances of 10 and 80 miles, such pigeons orient just as well as control birds (Fig. 17), and, even more surprising, some of them actually get home.

Schmidt-Koenig performed some of his experiments at my loft, and I therefore had the opportunity to see birds returning from the tests. It was truly a remarkable sight. The birds flew considerably higher than normal, and they did not swoop in for a landing on the loft like normal pigeons. Instead, they came almost straight down in a peculiar helicoptering or hovering flight. Being unable to see the loft itself, they landed in yards or fields in the vicinity, where we picked them up and carried them to the loft. Since we often failed to see birds land and only discovered them by chance in nearby weed patches and ditches, I am confident that additional birds suc-

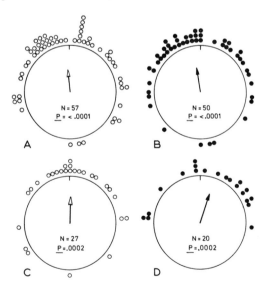

Fig. 17. Initial bearings at 10 miles (A,B) and 80 miles (C,D) of pigeons wearing contact lenses. The bearings of experimental birds (B,D) with frosted lenses are as well oriented as those of control birds (A,C) with clear lenses. (Note that the home direction is not indicated, because Schmidt-Koenig pools bearings by setting the control means of the various releases to zero, without regard for the home direction.) (After Schmidt-Koenig and Schlichte, 1972.)

ceeded in returning to within a few hundred yards of the loft but were never found. Thus it seems certain that the number of birds returning from all Schmidt-Koenig and Schlichte's tests was actually considerably higher than they are able to report.

These are very important experiments because they indicate not only that pigeons can orient with virtually no image-vision input, but also that pigeons returning from long flights can somehow tell when they are home even though they cannot see either the loft or the surrounding landmarks. If this is so, it strongly suggests that the pigeon navigation system is accurate enough to pinpoint home almost exactly—a degree of navigational accuracy unrivaled by even the most sophisticated of man's mechanical contrivances.

These results, of course, bring other questions immediately to mind. For example, do magnets interfere with the orientation of birds wearing the frosted lenses? Are the birds using the sun compass in their navigation? Can they orient under total overcast? Can inexperienced pigeons orient while wearing the lenses? Schmidt-Koenig (personal communication) is now conducting tests designed to answer such questions. His answers may well have an enormous influence on the course of future research on avian orientation and navigation.[15]

IV. THE "MAP" COMPONENT

We have seen that Kramer's (1953a) suggestion that homing birds must combine "map" information with compass information has proved useful in interpreting the results of many kinds of experiments. But concepts of the map have remained hazy, and the information on which such a map would have to be based has never been linked with any real environmental source. In my opinion, however, there are some possible clues that should receive greater attention. I shall mention two categories of these here—geographic and temporal deviations from the expected bearings.

A. DIFFERENCES IN ORIENTATIONAL PERFORMANCE AT DIFFERENT RELEASE SITES

Although the mean bearings of pigeons released at test sites usually point roughly homeward, they almost never point directly toward home. Moreover, at each release site the deviation of the mean direction from the true home direction is usually consistent for birds from a given loft. For example, Fig. 18A shows the mean vectors for twelve consecutive releases of the same

[15] Schmidt-Koenig and Walcott (1973) have recently published some preliminary results from airplane-tracking of pigeons wearing frosted lenses.

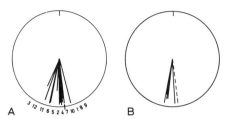

Fig. 18. Mean vectors of Cornell pigeons released at Weedsport, N.Y. (A) Twelve consecutive releases of the same birds; there is a clear tendency for the bearings to be slightly clockwise from the true home direction (only one mean is left of home). The release numbers for the means are given to enable inspection for any trend over the course of the twelve releases. (B) Means of six other groups of Cornell pigeons; the same bias is evident. (W.T. Keeton, unpublished data.)

group of pigeons at a site (Weedsport, New York) 45.6 miles north of our loft. It can be seen that the mean bearings usually deviated slightly clockwise from the home direction. That this was not due merely to some idiosyncrasy of the particular group of pigeons is established by Fig. 18B, which shows the mean vectors of six other releases, each utilizing a different group of pigeons; the same deviation is evident. Thus all seven groups of birds agree in departing to the right of the home direction. This kind of consistency usually holds from year to year, and from month to month within any given year. It is, in effect, a relatively stable characteristic of each release site.

Almost everyone who has worked on pigeon homing has found such release-site biases (Schmidt-Koenig, 1961, 1963a, 1965; Wallraff, 1959a, 1967, 1970b; Keeton, 1969). In homing experiments with bank swallows captured in colonies located near our lofts, we have found that these birds give initial bearings that usually show roughly the same deviations from the true home directions as our pigeons' bearings do at those sites (Fig. 19) (Downhower

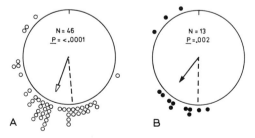

Fig. 19. A comparison of the initial bearings of Cornell pigeons (A) and bank swallows from a colony near the loft (B) released at Meridian, N.Y., 53 miles north. The two species show very nearly the same bias. (From J. F. Downhower and W. T. Keeton, unpublished data.)

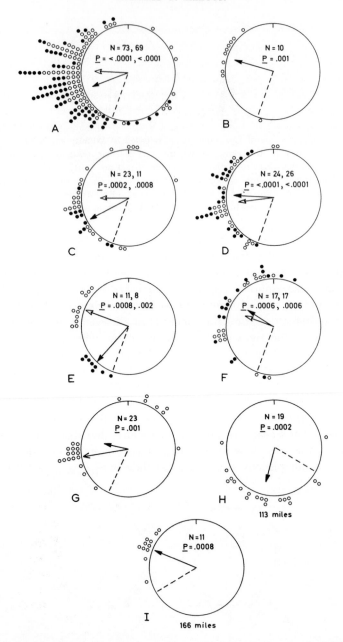

FIG. 20. Initial bearings for releases at Castor Hill, New York. (A) Pooled bearings from 13 releases of experienced Cornell pigeons under sun; open symbols, bearings of pigeons new to the site; solid symbols, pigeons previously flown from this site. (B) Bearings from a release of first-flight Cornell pigeons under sun. (C) Pooled

and Keeton, unpublished data). Thus release-site biases may not be species-specific.

It is my contention that release-site biases deserve more experimental attention, because they do not appear to be a function of the known or suspected compass mechanisms, nor is there any evidence that they are functions of the other possible orientational cues discussed above. Hence the intriguing possibility that they are related to the mysterious map information, as Kramer (1957, 1959) himself suggested, should not be overlooked.

1. Characterizing Release Sites

a. *The Castor Hill Site.* Though every release site's properties can be characterized if enough tests are done there, the sites that most readily lend themselves to detailed investigation are those where the deviation from the true home direction is quite large. One such site, the Castor Hill Fire Tower, 89 miles NNE of our Cornell lofts, has proved to be particularly interesting (Keeton, 1973). At this location our birds regularly depart with very little scatter but with mean bearings roughly 50°–90° clockwise from the homeward direction (for the results of a typical release, see Keeton, 1970c, Fig. 5B). Figure 20A shows the pooled bearings from 13 such releases in which normal unmanipulated pigeons with some previous homing experience were used; the clockwise bias is evident whether or not the birds had had previous releases from this site. First-flight pigeons, i.e., ones with no previous homing experience, choose similar bearings (Fig. 20B), so whatever it is that produces such a pronounced deviation at this site is not a function of experience. Moreover, releases under total overcast yield similar results (Fig. 20C), and so do releases under sun of pigeons wearing bar magnets on their backs (Fig. 20D) (unpublished data), so it seems unlikely that the orientational "error," if we may call it that for now, occurs in the compass step, whether

bearings from three releases of experienced Cornell pigeons (with and without previous flights from this site) under total overcast. (D) Pooled bearings from two releases under sun of experienced pigeons, new to the site, wearing brass bars (open symbols) or magnet bars (solid symbols) glued to their backs. (E) Bearings of control and 5-hours-fast clock-shifted pigeons under sun. (F) Bearings of pigeons wearing clear (open symbols) or frosted (solid symbols) contact lenses, under sun. (G) Bearings of bank swallows captured in a colony near the Cornell loft; the distribution appears to be bimodal, with the principal cluster in the same direction chosen by pigeons and the other in nearly the opposite direction. The vector with solid arrowhead and the values of N and P apply to the entire distribution; the vector with simple arrowhead is the mean of the principal cluster. (H) Pooled bearings from 2 releases of pigeons from Schenectady, New York. (I) Bearings from a release of pigeons from Fredonia, New York. [(D) Keeton, unpublished data. (F) Keeton and Schmidt-Koenig, unpublished data. Others from Keeton, 1973.]

the compass be sun or magnetic. This conclusion is reinforced by tests under sun in which clock-shifted pigeons give initial bearings that are deflected in the predicted manner from those of the control birds (*not* from the true home direction) (Fig. 20E), thus strongly suggesting that under such conditions the sun compass is functioning normally but that the "map" information is skewed. Vanishing bearings of pigeons wearing frosted contact lenses show the usual clockwise bias (Fig. 20F), so the biasing factor (the map?) must not depend on image vision (Keeton and Schmidt-Koenig, unpublished data).

When bank swallows from Ithaca are released at Castor Hill, most depart in the same direction as the pigeons, except that a few select the back azimuth (Fig. 20G). Apparently, then, these two different species agree that to get to Ithaca from Castor Hill the way to begin is to fly westward.

Several investigators have reported that birds from different lofts located only a few miles apart sometimes exhibit different deviations at a given release site (e.g., Wallraff, 1967, comparing Nordenham and Wilhelmshaven lofts; Schmidt-Koenig, 1963a, comparing four neighboring lofts in North Carolina). However, we have used pigeons from four other lofts in releases at Castor Hill, and have found that they perform quite like our Cornell birds. Two of these lofts are located in or near Ithaca, within 6 miles of our Cornell lofts, and pigeons from them choose initial bearings similar to those of Cornell pigeons. The third loft is in Schenectady, New York, which is approximately 150 miles east of Ithaca. When released at Castor Hill, pigeons from this loft deviate clockwise from their true home direction by an amount (70°–79°) similar to the deviation of Cornell pigeons from their home direction (Fig. 20H). Likewise, pigeons from the fourth loft, which is in Fredonia, New York, approximately 150 miles west of Ithaca, deviate clockwise from their true home direction by a similar (53°) amount (Fig. 20I).

Since pigeons from five different lofts and three very different destinations, as well as two different species of birds, appear to make roughly the same "error" at Castor Hill, and since the same "error" is apparent under both sun and overcast,[16] and whether or not the birds are experienced homers, I tentatively conclude that some environmental factor basic to the avian homing process is rotated clockwise at Castor Hill. This factor, whatever it may be, must be at least a part of what we have been calling the map.[17]

---

[16] Our evidence indicates that release-site biases are the same under both sun and overcast at other sites also.

[17] I must make it clear that my use of the term "map" throughout this section is not meant to imply anything about the nature of the cue or cues involved, or even that the "map" component of the orientation process comprises a single step. I use the term only as a convenient way of designating all information,

If this is so, then the birds are not making an error when they depart in a direction different from the one we might have anticipated. They are reading the map cues correctly, but the map is twisted. Indeed, we have evidence that to take up an initial bearing straight toward home is biologically in error, because pigeons clock-shifted 5 hours fast and thus aimed toward home have much poorer homing success than controls that depart in the usual westward direction. In short, it appears that if one were faced with the task of navigating from Castor Hill to Ithaca by the pigeon system, one properly should depart westward.

The question that now confronts us is what might the rotated environmental factor at Castor Hill be? We have thus far found no convincing answer, but I remain convinced that here is a clue of fundamental importance if we could only decipher it.

b. The Jersey Hill Site. Equally as interesting as sites where deviant bearings are regularly obtained are sites where pigeons have great difficulty orienting. We have studied such a site, the Jersey Hill Fire Tower, located 74.5 miles west of Ithaca (Keeton, 1971b). Schmidt-Koenig (1971, p. 338) has mentioned that sites of disorientation also exist near three of the localities (Wilhelmshaven, Frankfurt, Durham) where he has worked.

We have made numerous test releases at Jersey Hill, and in all but one instance the birds have departed randomly or nearly so (see Keeton, 1970c, Fig. 5A, for a typical release of pigeons never previously at the site), and the homing success has been poor. Even if the birds that return home are taken back to this site a second or third time, their initial bearings remain random (Fig. 21A), i.e., experience at the site brings no improvement in the bearings. We have tested pigeons from other lofts, and they too are ordinarily disoriented at Jersey Hill. Airplane tracking (courtesy of Charles Walcott) of four pigeons revealed that the birds remained "confused" (i.e., flying apparently aimlessly about in the general area of the release site) for at least 45 minutes.

One release at Jersey Hill deserves special mention. On 13 August, 1969, we tested pigeons from both Ithaca and Fredonia, which lie in nearly opposite directions from Jersey Hill. On this one day, our Cornell birds vanished in a tight cluster to the northeast, and the Fredonia birds vanished nonrandomly westward (Fig. 21B). We were never able to discover anything unusual about 13 August that could explain why the pigeons' behavior was so different from usual.

---

other than simple compass information, that enables a bird to determine the direction it should fly to get home. Whether such map information permits a bird to ascertain its position relative to home in both direction and distance or only direction is not clear.

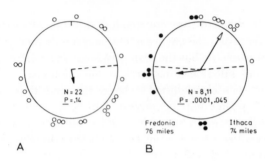

Fig. 21. Releases at Jersey Hill, N.Y. (A) Bearings from a release of Cornell pigeons that had been released at this same site twice previously. (B) Bearings of Cornell and Fredonia pigeons released on August 13, 1969. Both groups vanished nonrandomly. (From Keeton, unpublished data.)

Our efforts to learn what environmental attribute(s) of Jersey Hill might be so disorienting to pigeons have so far been unsuccessful, though some interesting anecdotal information has turned up (Keeton, 1971b). We have recently established two lofts in the Jersey Hill area (west of Hornell, New York), and hope soon to learn whether pigeons can home to this area even if they can't home from it.

2. The Role of Local Topography in Release-Site Biases

A thought that may come to the reader's mind when biases characterizing individual release sites are being discussed is the possibility that they may be due to local topographic features. Perhaps, for example, the consistent slight clockwise deviation of the mean bearings from the true home direction at Weedsport (Fig. 18) indicates that there is some feature of the landscape in that direction that is especially attractive to the pigeons. And perhaps at Castor Hill there is a similarly attractive feature to the west, or an intimidating one toward Ithaca.

Though it is true that really major topographic features, such as the seacoast, large lakes, high mountains, may influence initial bearings (e.g., Kramer, 1957; Wagner, 1968, 1970, 1972a), the more usual landscape variations, such as those seen at Weedsport and Castor Hill, appear to have little or no effect on pigeons' choice of directions. That this is so is indicated by the kinds of experiments we have conducted at Castor Hill. Thus, for example, we could aim the pigeons in virtually any direction desired simply by giving them an appropriate clock shift. Clearly, then, they are not merely flying toward some attractive hill, valley, or other landscape feature. Note that the effect of the clock shifts, at Castor Hill and all other sites we have studied, is to deflect the initial bearings by approximately the predicted amount from the bearings of the control birds, not from the home bearing (Fig. 20E). This indicates that the bearings of the controls are the result of the birds' processing of navigational information and not simply a tactic response to a topographic feature. Moreover, the fact that pigeons from the Schenectady loft vanished in a direction that nearly coincided with the true home direction of the Cornell pigeons (Fig. 20H) indicates that unmanipulated pigeons, not just clock-shifted ones, are quite willing to fly in that direction if their navigational information indicates they should—there is obviously no intimidating topographic feature that prevents pigeons from departing straight toward Ithaca. Finally, and most convincing, roughly the same bias is shown at Castor Hill by pigeons that are prevented from seeing the landscape by frosted contact lenses (Fig. 20F).

I therefore consider it reasonable to suppose that many of the release-site biases are due, at least in part, to peculiarities of the map information birds obtain there, and that working out the nature of these peculiarities is a major challenge for the future.

3. Is There a Pattern Discernible in the Orientation of Pigeons at Different Release Sites?

a. Possible Geographic Patterns. If release-site biases are partly a function of the map information available at the sites, then it is important to determine whether there is any geographic pattern to the biases, i.e., whether they are strictly the result of local peculiarities or whether they reflect more widespread irregularities.

A first point to be made is that two sites very close to each other may be characterized by quite different biases. For example, one of my students, Donald M. Windsor (1972), conducted a series of simultaneous releases of first-flight pigeons at two sites, north of the loft, that are only 6 miles apart. At one of these sites, first-flight birds regularly deviate markedly from the true home direction, usually vanishing in the northwest. We have never found a consistent bias at the other site; means are sometimes slightly clockwise from home and sometimes slightly counterclockwise, with the mean of means directed almost straight home. As Fig. 22 shows, the results of Windsor's simultaneous releases are consistent with this pattern, indicating that the markedly different behavior of the birds at these two neighboring sites cannot be explained by differences between days, times of day, weather conditions, direction from the loft, or previous experience.

Despite such curious differences between neighboring release sites, however, there does tend to be a rough overall pattern to the biases exhibited by our Cornell birds. Windsor (1972) has found that at sites east of a line running NNW–SSE through the loft our birds tend to choose initial bearings that deviate clockwise from the true home direction, whereas west of that

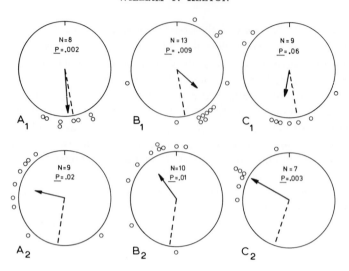

FIG. 22. Simultaneous releases of first-flight pigeons at sites 17 miles (top row) and 11 miles (bottom row) north of the Cornell loft on May 5 (A), May 7 (B), and May 11 (C), 1971. The biases of first-flight birds at the two sites are very different. (From Windsor, 1972.)

line they tend to deviate counterclockwise. Windsor interprets this as indicating that our birds have a directional preference for the northwest, but this is not completely compatible with the fact that along the dividing line, whether NNW of the loft or SSE, the birds are oriented with very little scatter almost straight toward home; i.e., the bearings give little indication that orienting SSE is any more difficult for them than orienting NNW.

Windsor (1972) has applied the same kind of analysis to the unpublished mean vectors obtained by Schmidt-Koenig (1966, 1970) in systematic studies of his birds' orientation at sites around Durham, North Carolina, and Frankfurt, West Germany.[18] In general, the Frankfurt sites appear to segregate along a line running roughly E–W through the loft, with the biases north of the line being clockwise and those south of the line counterclockwise. No clear pattern is discernible for the Durham sites. Nonetheless, Windsor interprets both sets of Schmidt-Koenig's data, as well as his own data and the data of Graue (1970), Kramer (1957), Kramer et al. (1957, 1958), Pratt and Wallraff (1958), Wallraff (1959a, 1967, 1970b), and Papi et al. (1971b) as indicating that there is a widespread tendency for pigeons to vanish preferentially toward the northwest and thus to have greater homing success when released in the south or southeast. He considers this to be con-

[18] I thank Dr. Klaus Schmidt-Koenig for so generously making these unpublished data available to us. They were of great value to Dr. Windsor in his study.

sistent with the widespread tendency of waterfowl to adopt so-called "non-sense" orientation to the northwest, as discussed by Matthews (1961, 1968). He admits, however, that tendencies toward other directions have sometimes been found in pigeons (e.g., Wallraff, 1967, 1970b, for inexperienced birds at Wilhelmshaven; Graue and Pratt, 1959, for Sacramento, California and Cedar Rapids, Iowa; Hoffmann, 1959, for Cambridge, England) and in several other species, among them Canada geese (Bellrose, 1963), green-winged teal (Matthews et al., 1963), common terns (Griffin and Goldsmith, 1955), and herring and ring-billed gulls (Southern, 1969a,b,c).

Two different interpretations for the kinds of biases we have been examining here have been proposed. Kramer (1959) called attention to "the manifold deviations from the true home-direction" found at different release sites, and suggested that, in northwestern Germany, "strictly localized troughs and peaks exist in what may be called a relief map of orientation facilities." In short, he attributed the biases partly to differences in the environmental information available to the birds at the different sites. By contrast, Wallraff (1967, 1970b) appears to put more emphasis on the home location itself and its effect on the behavior of the birds, i.e., he assumes that something about the loft imposes a strong directional preference on the pigeons, and that at release sites the birds choose initial bearings that are a compromise between the loft-specific preferred direction and the home direction. He puts less emphasis than Kramer on possible distortions of map information at the release sites. It should be obvious from what I have said already that I favor Kramer's point of view.

b. *Are There Patterns Due to Distance Effect?* Several investigators have claimed that a pattern is evident when initial bearings are analyzed in terms of the distances of release sites from the home loft. Matthews (1955a, 1963b) reported that his pigeons exhibited poorer orientation at intermediate distances than at very short or long distances. He interpreted this "distance effect" as indicating that the birds pilot by familiar landmarks at short distances and use their true navigation system at long distances, but that at intermediate distances the birds are too far from home to use landmarks and too close to use accurately their bicoordinate navigation system. Schmidt-Koenig conducted extensive tests of the distance effect in both North Carolina (Schmidt-Koenig, 1964, 1966) and Germany (Schmidt-Koenig, 1968, 1970), and published results that seemed to support Matthews' ideas. Indeed, Matthews (1968) has relied heavily on Schmidt-Koenig's work in reinterpreting much of the published data on pigeon homing that seems to conflict with his sun-arc hypothesis. Wallraff (1967, 1970b) has also reported finding a distance effect, but Graue (1970), Keeton (1970c), and Windsor (1972) have not found the effect in their birds.

Figure 23 (top) shows the curves of distance plotted against homeward

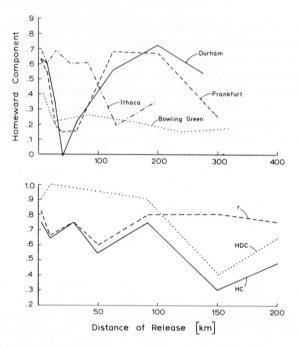

FIG. 23. (Top) Graphs of homeward component against distance obtained by Schmidt-Koenig (1966, 1970) at Durham, N.C., and Frankfurt, West Germany; by Graue (1970) at Bowling Green, Ohio; and by Keeton (1970c) at Ithaca, N.Y. (Bottom) Keeton's (1970c) curves for length of mean vector (r), homeward directional component (HDC), and homeward component (HC) for releases from the north.

component[19] obtained by Schmidt-Koenig (1966, 1970) at Durham and Frankfurt, by Graue (1970) at Bowling Green, Ohio, and by Keeton (1970c) at Ithaca, New York. Though Schmidt-Koenig's two curves are remarkably similar, they show little resemblance to the curves of Graue and Keeton, which are, in turn, unlike each other.

I have pointed out elsewhere (Keeton, 1970c) that curves such as those shown in Fig. 23 (top) can be misleading for two reasons: (1) Each of these is a summary curve calculated by combining data obtained from a series of releases in each of the four cardinal compass directions, but the separate curves for the four directions usually show little resemblance to each other. Hence the summary curves are, in a sense, artifacts of the experimenter's compulsion to reduce four curves to one. (2) The homeward com-

---

[19] The homeward component, h, is the projection of the mean vector onto the home vector, calculated as $h = a \cdot \cos(\alpha - \beta)$, where a is the length of the mean vector, α is the mean direction, and β is the home direction.

ponent is a function of two quantities—the deviation of the mean direction from the home direction and the degree of scatter of the bearings—that do not vary concurrently. Hence a homeward component of zero can result from any of a large number of very different distributions of vanishing bearings, including two extreme ones that are as fundamentally different as it is possible for two distributions to be—a circularly uniform distribution (i.e., one with no mean vector) or a distribution in which all bearings deviate 90° in the same direction from home (i.e., one for which the mean vector is of maximum length but is oriented at right angles to home).

I argue that accuracy of the mean direction and extent of scatter are two very different things biologically, yet the homeward component permits no distinction between them. Much better, in my opinion, is a graph that shows both the length of the mean vector (r), as an inverse measure of scatter, and the homeward directional component (HDC), calculated by the equation $d = \cos (\alpha - \beta)$, which is a function only of the deviation of the mean direction from the home direction. These two indices give independent measures of the two variables that together determine the homeward component. Figure 23 (bottom) shows the curves for r and HDC for my distance-effect tests from the north, together with the curve for the homeward component (HC) that they determine. It can be seen clearly that the shape of the HC curve at the shorter distances is due almost entirely to variations from site to site in the degree of scatter of the bearings, the accuracy of the mean direction changing only slightly, whereas at the longer distances the shape of the curve is determined by variations in the accuracy of the mean direction, the extent of scatter remaining nearly constant.

I conclude that the distance effect is not a universal characteristic of pigeon homing, and, furthermore, that r and HDC reveal much more than HC about what is happening at each release site, yet neither of these quantities varies consistently with the distance of the release site from the loft.

4. Can Bearings Be Altered by Training?

Virtually all investigators who have worked with pigeons have agreed that homing success (i.e., homing speed and percentage of birds returning) improves with experience (e.g., Matthews, 1953b; Hoffmann, 1959; Wallraff, 1959b; Schmidt-Koenig, 1963b; Sonnberg and Schmidt-Koenig, 1970), though the most important thing is not experience at the particular test sites, or even the distances flown, but the number of homing flights (Wallraff, 1959b; Keeton, unpublished data). However, there has been disagreement on whether there is a similar improvement in initial bearings. Thus Schmidt-Koenig (1965) reports a reduction in scatter as a result of experience, whereas several other authors (e.g., Matthews, 1953b; Pratt and Wallraff, 1958; Wallraff, 1959b) find no such improvement.

When discussing the influence of experience, it is important to distinguish between the first few homing flights and later flights. Thus we have found that, when conditions are optimum, first-flight pigeons frequently give initial bearings that are no more scattered than those of very experienced birds at the same site on the same day, but that such pigeons are much more susceptible to being disoriented by various disturbing factors, such as lack of sun (Keeton and Gobert, 1970), attached magnets (Keeton, 1971a, 1972), attempted clock shifts (Alexander, 1974), or unsettled or cold weather (Keeton, unpublished data). Pigeons that have had only two or three short-distance training flights often respond to such conditions more like very experienced birds than like first-flight ones. Thus it may well be that only the first few flights significantly affect the scatter of bearings—and then only under certain conditions. The investigators who found no reduced scatter with experience may have happened to conduct their tests when there were no disturbing conditions, or they may have failed to distinguish between the first few flights and later training flights.

More important than changes in scatter to our discussion of the map component is the effect of experience on the mean directions chosen. I am convinced that first-flight pigeons often differ from experienced birds in their reading of map information. Thus at some release sites first-flight pigeons consistently show biases significantly different from those of more experienced birds. For example, at the Weedsport site discussed earlier (Fig. 18), where experienced birds usually deviate slightly clockwise from the true home direction, first-flight birds regularly deviate counterclockwise. The situation is similar at another site near Weedsport; experienced birds regularly vanish clockwise from the home direction, whereas first-flight birds vanish counterclockwise, as shown in Fig. 6C. Figure 24 shows the effect of a single previous release on the bias at a site 11 miles north of the loft, where first-flight pigeons consistently depart to the northwest (see Fig. 22, lower row); second-flight birds vanish to the southwest, like experienced birds, whether their first flight was from this particular site (Fig. 24A) or from a site only 3.7 miles north of the loft (Fig. 24B). Wallraff (1959), too, has found systematic shifts of release-site bias with experience. His shifts have sometimes been attributed to an asymmetric pattern of training releases, but I disagree with this interpretation since his results appear to be similar to mine, and mine cannot be explained in this way.

I don't mean to imply that the biases of first-flight and experienced pigeons differ significantly at all release sites, because they certainly do not. For example, at Castor Hill first-flight birds choose initial bearings (Fig. 20B) that are similar to those of experienced birds. But the fact that significant differences between the two categories of birds do occur at some release sites leads me to think that the first few homing flights must have some

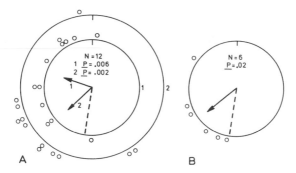

F<small>IG</small>. 24. Effect of a single previous release on the bias at the 11-mile site shown in Fig. 22, bottom row. (A) Bearings of first-flight pigeons at this site (inner circle) and of the same birds on their second release (outer circle). (B) Bearings of second-flight birds whose first flight was from a site 3.7 miles north of the loft. (From Windsor, 1972.)

important maturational effect on utilization of map information. Hence this may prove to be a useful tool in trying to probe the nature of the map.

In addition to the effects of experience discussed so far, which depend very little on the direction of the training flights, many investigators (Riviere, 1929; Kramer and von St. Paul, 1950; Matthews, 1951b; Hitchcock, 1952; van Riper and Kalmbach, 1952; Pratt and Thouless, 1955; Michener and Walcott, 1966, 1967a; Walcott and Michener, 1967; Wallraff, 1959b; Baldaccini *et al.*, 1971) have reported that pigeons trained from a single direction tend, upon release at a site off the training line, to take up initial bearings in the training direction, even though that direction is no longer appropriate. Graue (1965) and Wallraff (1967) have even reported that a single flight will influence initial orientation on the succeeding release, and Graue suggests that the effects of several releases are averaged to influence the orientation on later flights. However, other tests (e.g., by Heinroth and Heinroth, 1941; Matthews, 1951b; Alexander and Keeton, 1972; Walcott, 1972a) have failed to reveal such a training bias.

Many of the experiments on training bias have been conducted with very small samples or without adequate controls, as discussed by Schmidt-Koenig (1965), Alexander and Keeton (1972), and Alexander (1970). (The latter author provided a detailed reanalysis of all data published on the subject up to that time.) But even if we discount many of the papers for the reasons mentioned, we are still left with a discrepancy—of the two major investigations on this topic, Graue's results, taken as a whole (the means in individual tests were seldom statistically different), seem to support the idea that each release exerts a biasing effect on the next release, and Alexander and Keeton's results do not. Thus no final resolution of the debate is yet possible, but I conclude that if such an effect exists, it is not a general phenomenon and is probably of minor importance for pigeons trained from all directions.

B. Temporal Changes in Orientation and Homing Success

I have said that release-site biases tend to be quite stable, and this is true if one looks at the overall picture. But a universal observation of those who work with pigeons is that there are many variations, some minor and some major, in the birds' behavior. These variations may be in the extent of scatter, the mean direction chosen, the elapsed time between release of the birds and their departure, the homing speeds, or the percentage of birds returning. Most such variations remain unexplained.

An important, but usually difficult, point is to distinguish between variations that are due to changes in the birds' health, condition, or general motivation (which may be influenced by the stage of the reproductive cycle, the degree of hunger or thirst, etc.) and those that are due to altered environmental conditions. It is with the latter that we shall be concerned here.

Many investigators (e.g., Wallraff, 1959a) have observed marked changes in orientation during the course of a day, even when no corresponding changes in the weather are obvious. In an attempt to determine whether there is any consistent change with time of day, I have calculated the Spearman rank-correlation coefficient for time of release versus deviation from the home bearing, deviation from the mean bearing, and homing speed for hundreds of test releases, and find that no statistical correlation exists for any of these three pairs of variables; the coefficient has a random probability of 0.05 or lower in only approximately 5% of the tests, which is, of course, what is predicted on the basis of chance alone. In a different approach to the same question, Windsor (1972) used the same 12 preselected times of day for releases on 12 different days, and plotted the means thus obtained against time (Fig. 25); no pattern of change was evident. I conclude, then, that the time of release has no consistent relationship to either the accuracy of the initial bearings or the homing speed (except in extreme cases, as when releases are conducted very late in the day). Thus it appears that intradiurnal changes in these variables are aperiodic. Wallraff (1971) has suggested, in fact, that some aperiodic fluctuations in environmental factors influencing pigeon orientation probably last, on the average, less than 10 minutes; he finds that closely similar vanishing bearings from pigeons released within 5 minutes of each other (but with no opportunity to influence each other during takeoff) are more frequent than can be expected by chance.

There are occasional dramatic changes from one day to the next in the homing behavior of the same, or similarly experienced, pigeons released at

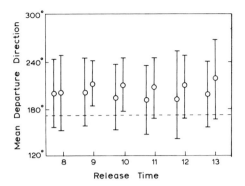

Fig. 25. Departure bearings at 12 times of day at a release site 46 miles north of the loft. Each symbol shows the mean and angular deviation for twelve different birds released on 12 different days. The dashed line indicates the home direction. No consistent correlation of bearings with release time is evident. (From Windsor, 1972.)

the same site. In some instances, the change may be from tightly clumped, well-oriented bearings on one day to widely scattered, or even random, bearings on the next, or the reverse. In other instances, the change is a marked shift in the mean direction chosen. In still other cases, the change may be from fast homing speeds to very slow ones. I think it likely that at least some of these major changes in orientation behavior are related in some unknown way to changes in map information, as Kramer (1959) suggested.

A few actual examples of such day-to-day changes in orientation may be useful here. One has already been mentioned—the single day on which two independent samples of pigeons gave nonrandom departure bearings at a site where on other test days the bearings were random (Fig. 21B). Another two-sample example occurred at a release site 21 miles east of our loft. In the more than 70 other test releases we have conducted at this site (both before and since), the mean bearings of normal birds, and those of control birds in clock-shift tests, have regularly been well oriented homeward (usually slightly north of the true home direction), and the mean bearings of birds clock-shifted 6 hours fast have been southward-directed (see Keeton, 1969, Figs. 2–6). Yet on this particular day the control birds vanished nonrandomly northward (353°) and the clock-shifted birds westward (266°). In other words, the means of the two treatments differed by roughly the amount expected, but both were rotated nearly 90° clockwise from the predicted directions. It was as though the map itself had been rotated, affecting both treatments in the same way. In neither of these examples were we ever able to discover anything unusual about the days on which they occurred. For other examples, see Wallraff (1959a).

Wallraff (1959a, 1960a) has examined in detail possible correlations of the several measurable parameters of pigeon homing behavior with various meteorological and geophysical factors such as wind, visibility, air temperature, barometric pressure, barometric pressure changes, sunspot activity, and geomagnetic disturbances. In most cases he found no correlation (except ones he considered "trivial," such as poor homing in dense fog). However, he did note a significant inverse correlation between homing performance and interdiurnal barometric pressure changes in the middle and upper troposphere, and also between mean bearings and these same pressure changes. Both Wallraff (1959a) and Kramer (1959) felt the correlations are based on qualitative changes in the navigational map of the pigeons.

Since no such correlations with barometric pressure changes could be seen at the altitudes pigeons fly, the biological relevance of these correlations has remained in question, and little further attention has been paid to them. However, recent experiments (Kreithen and Keeton, 1974a) indicate that pigeons possess an extraordinary sensitivity to pressure changes (Fig. 26), hence it is possible that changes near ground level smaller than those measured by Wallraff influence homing behavior. This sensitivity to pressure might also act as an altimeter for birds flying within dense clouds, and it might help explain the remarkable ability of many migrants to predict meteorological conditions, especially winds aloft, in choosing which nights to initiate migratory flights (Emlen, in preparation).

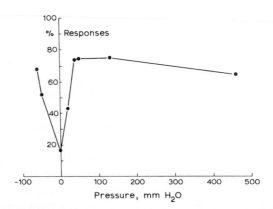

Fig. 26. Sensitivity of a pigeon to small changes in barometric pressure. The bird has been conditioned to pressure changes (the amount indicated, per 5 seconds) as signals of an impending electric shock. The response is an elevated heart rate. This bird gave responses significantly above the background level to changes of ±10 mm H$_2$O which is equivalent to an altitude change of approximately 33 feet. Some birds could detect pressure changes equivalent to altitude changes of less than 20 feet. (From Kreithen and Keeton, 1974a.)

Wallraff (1959a, 1960a) also documented an annual periodicity in homing performance (but not in initial bearings) that had been suggested earlier by Kramer (1954, 1957) and Kramer and von St. Paul (1956). He reported that this cycle roughly parallels the annual cycle of daily mean temperatures, with homing being best in August and worst in January. But, according to Wallraff (1960a; see also Kramer, 1959), short-term (aperiodic) fluctuations in homing performance and temperature are correlated only during the astronomical winter (September–March), not during the summer (April–August) when homing is best. Wallraff could find no evidence that the annual cycle was related to the breeding cycle, and both he (Wallraff, 1960a) and Kramer (1959) felt that it probably does not indicate a direct effect of mean temperature, *per se,* on homing, but rather that there are seasonal qualitative changes in the navigational map of the pigeons.

Gronau and Schmidt-Koenig (1970) have recently reported that they too find an annual cycle of homing performance, but they, unlike Wallraff, also report a corresponding cycle in the accuracy of initial bearings (analyzed in terms of the homeward component). Their orientation data for naïve birds look reasonably convincing, but the cyclic nature of the data for experienced birds is not so evident. Only if data for several different years show essentially the same curve, when graphed separately, can a truly convincing claim be made that there is a regular yearly cycle of orientational accuracy.

One last type of orientational change, which is very likely an indication of altered map information, must be mentioned, even though no explanation for it can yet be given. This is a rather sudden and long-lasting change in the characteristic departure bias of a release site. Several such changes have been reported. For example, Wallraff (1959a, Fig. 22) discusses a site near Wilhelmshaven where there was a clear tendency for his birds to vanish to the southeast during the years 1954–1956, and a similarly clear tendency for them to vanish more northeasterly during the years 1957–1958. It is to be hoped that if other such bias shifts occur in the future we will be able to utilize them fully in our attempt to probe the nature of the elusive navigational map.

V. Some Comments about Techniques

Initial bearings, obtained visually, have traditionally been emphasized in experiments on homing, and we have seen in the present paper that they provide much valuable information about the earliest stage of the homing process. But they also have several shortcomings: (1) There is usually considerable "noise" in the data at even carefully selected release sites due to loss of birds behind distant trees or other obstacles. Most irritating to the observer is the not infrequent situation where a bird must be recorded as lost from view at a bearing that clearly does not reflect the direction in which it actually was traveling—for example, when a bird considerably east of the observer is flying from north to south but is inadvertently lost from sight at 250°. (2) Initial bearings alone cannot help us explain why, at sites like Castor Hill, birds that depart at a considerable deviation from the true home direction nonetheless have better homing success than clock-shifted birds that depart straight toward home. In other words, we need to know more about reorientation along the way, more about the entire flight path from release site to home. (3) It is very difficult to manipulate some of the possible orientational cues when one must work with free-flying birds. Let us examine each of these points in turn.

a. *Visual Bearings.* Even though visually obtained initial bearings are noisy data, several investigators (e.g., Matthews, 1951b; Wallraff, 1959a) have concluded that there is a positive correlation between the homeward accuracy of the bearings and homing performance. Wallraff (1959a) claims this is true whether one is considering individual differences in a single test, or differences in the mean vectors at different release sites, or variations in the mean from day to day at the same site.

It is certainly true that in 6-hour clock-shift tests under sun the marked separation of the departure bearings of control and experimental birds is correlated with dramatic differences in homing success (Schmidt-Koenig, 1958, 1960, 1961; Keeton, 1969). It is also true that at sites like Jersey Hill, where bearings are random, homing success is similarly poor. But I do not find a consistent correlation between bearings and homing success when I compare ordinary releases at sites where the biases are not so extreme. Nor do I find a correlation when I compare the individual birds within a single test.

Let us look more closely at the last point. My associates and I have conducted considerably more than 1,000 test releases utilizing more than 10,000 single-tossed pigeons. The data from each test have been analyzed by computer. A routine part of this analysis is calculation of the Spearman rank-correlation coefficient for the following pairs of variables:

1. deviation from the home direction—homing speed
2. deviation from the home direction—corrected homing speed[20]
3. deviation from the mean bearing—homing speed
4. deviation from the mean bearing—corrected homing speed

---

[20] Homing speed is based on elapsed time between *release* and arrival at the loft. Corrected homing speed is based on elapsed time between *vanishing* from the release site and arrival at the loft, with a slight adjustment of the distance according to the direction of vanishing.

5. vanishing interval—homing speed
6. vanishing interval—corrected homing speed
7. vanishing interval—deviation from the home direction
8. vanishing interval—deviation from the mean bearing
9. time of release—deviation from the home direction
10. time of release—deviation from the mean bearing
11. time of release—vanishing interval
12. time of release—homing speed

If we look only at tests involving unmanipulated pigeons (i.e., ones not clock-shifted or wearing magnets, etc.) we find that the calculated coefficients for each of the 12 pairs of variables have a random probability of 0.05 or lower in roughly the percentage of tests that would be expected on the basis of chance. However, if we perform a second-order analysis for each pair of variables, by applying the sign test (two-tailed) to the coefficients from hundreds of tests, we find that there are significantly more negative than positive coefficients for pairs 5 ($P = 0.004$), 6 ($P = 0.002$), and 11 ($P = 0.004$). In short, there is no indication at all of a correlation between bearings and homing speeds, contrary to what Wallraff reported,[21] but there appears to be a weak inverse correlation between vanishing interval and homing speeds—birds that take less time before departing tend to home faster. The third inverse correlation, that between time of release and vanishing interval, may indicate that the birds obtain some navigational information while sitting in the baskets at the release site, or it may indicate simply that the birds that have waited longer in the baskets are tireder, hungrier, and thirstier and hence more motivated to depart quickly for home.

It may be a bit disconcerting that we find no correlation between bearings and homing speeds within single tests, but it is not really too surprising. In tests where the bearings are nonrandom, most birds depart within somewhat less then half the circle, i.e., relatively few birds take up wildly different bearings. Moreover, part of the scatter in visual bearings is due to the "noise" in the data mentioned earlier, i.e., birds recorded as having departed in a direction in which they were not actually flying. If we could distinguish the birds that kept going in odd directions, we might very well find that they do take longer to home.

b. Radio Bearings. One way of following the birds longer, and hence of eliminating some of the noise in the data, is to use radio telemetry. Small transmitters attached to the birds permit observers to track them until they are 4 to 10 miles from the release site. We have found that at most sites

---

[21] The table published by Wallraff (1959a) combines data from comparisons within single releases, between releases at the same site, and betweeen releases at different sites, hence it is impossible to evaluate his within-release data separately.

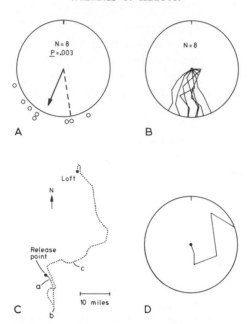

F‍IG. 27. A comparison of visual bearings, radio-tracking data, and airplane-tracking data. (A) Visual bearings from a release north of the Cornell loft. (B) Radio data for the same release; each line connects five bearings for a single bird: the bearings after 20%, 40%, 60%, 80%, and 100% of the tracking time. Note that the final bearings for the 8 birds are more tightly clustered and more accurately homeward-oriented than the visual bearings. (C) An entire flight path mapped by airplane-tracking (Walcott and Michener, 1971). This rather atypical path has been chosen for illustration in order to indicate how different visual, radio, and airplane data from the same flight may sometimes be. Thus this bird would probably have been lost visually at about point *a*, giving a southwest bearing. If we assume radio tracking to have been possible up to 10 miles from the release point, then the final radio contact might have been at point *b*, or, with luck, at point *c*, giving southerly or east-northeasterly bearings. (D) Radio data for the same flight, from release to point *c*.

the principal difference between visual and radio bearings is that the scatter is reduced in the latter (compare Figs. 27A and 27B). However, at some locations, where visual techniques indicate the birds depart in a direction significantly different from the true home direction, radio tracking shows that they soon begin to turn more homeward (Fig. 28). Thus radio tracking from the ground provides more information than visual tracking because more of the homing flight can be monitored. But it still leaves many questions unanswered. For example, it does not help us explain our results from Castor Hill, because at that site the final radio bearings of our birds are still westward, like the visual ones.

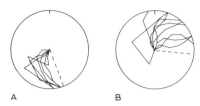

Fig. 28. Radio-tracking data for releases 5 miles north (A) and 30 miles west (B). It is evident that the pigeons began to correct their courses and turn onto a more homeward-oriented path soon after they would have been lost from visual contact. (From Windsor, 1972.)

c. Airplane Tracking. The entire flight home can sometimes be monitored if a bird is tracked by airplane (Fig. 27C) or helicopter. This technique was used quite effectively by Griffin (1943, 1952b; Griffin and Hock, 1949) and Hitchcock (1952, 1955), both of whom followed the birds visually. But there are several difficulties in using visual tracking from winged aircraft: First, even if the aircraft is a very slow-flying one, it is very difficult to maintain contact with the birds, especially with single pigeons, hence flocks must ordinarily be used. Though there is evidence that small flocks do not differ significantly from single pigeons in either initial bearings or homing speeds (Keeton, 1970d), it is nonetheless usually preferable to use single-tossed birds in experiments. Second, most modern winged aircraft cannot be flown slow enough to stay with a homing bird. Third, there is some indication (Michener and Walcott, 1966) that pigeons are disturbed and show abnormal behavior if such an aircraft comes within $\frac{1}{4}$ mile of them. For these reasons, Michener and Walcott (1966, 1967a; Walcott and Michener, 1967) developed a technique for radio tracking pigeons from airplanes. They have succeeded in following hundreds of pigeon flights in this way.

An alternative solution to the problems mentioned above has been pursued by Wagner (1970, 1972a). He uses a helicopter to follow pigeon flocks visually. Because of its superb maneuverability and speed control, a helicopter is admirably suited for such a task. Moreover, unlike winged aircraft, a helicopter apparently causes no disturbance to the birds, even when it approaches as close as 30 meters; Wagner's report on this point has been independently confirmed by Talkington (personal communication) and by Walcott (personal communication).

The airplane and helicopter studies have made it possible to answer several important questions concerning events during the homeward flight:

(1) Do birds homing from a familiar site follow the same course each time? The answer is no. Michener and Walcott find that pigeons almost never follow the same route. Furthermore, when they encounter a "familiar"

course while homing from a new site, they give no indication that they recognize it.

(2) Is there evidence of reorientation en route? The answer is yes. Michener and Walcott saw numerous cases of major changes of course during flight, and many of these changes were unrelated to topography. That reorientation en route must occur was, of course, evident from ground studies. Otherwise it would be inexplicable that birds departing from a release site in a direction significantly different from the home direction, as at Castor Hill, often get home in good time (airplane tracking at Castor Hill, courtesy of Charles Walcott, indicates that a turn onto a southward course occurs roughly 14–18 miles from the release site).

Results of some of my clock-shift tests also provide strong evidence for reorientation en route (Keeton, unpublished data). Thus there are three cases in which clock-shifted birds released under total overcast gave homeward-directed initial bearings like those of the controls, but, unlike the results of most such overcast tests, the clock-shifted birds had much poorer homing success than the controls. In all three instances, the birds had encountered sunny conditions part way home. It seems highly probable that the clock-shifted birds reoriented using their sun compass, and consequently veered off the homeward course. It would be most interesting to document such an event by airplane tracking.

(3) Do topographic features encountered en route greatly influence the flight path of pigeons? Wagner's (1968, 1970, 1972) studies in the Swiss Alps provide dramatic answers to this question. He finds that large lakes or deep valleys seriously influence initial bearings recorded visually from the ground, because pigeons, when first released, generally avoid flying over the water and prefer flying along valley axes. But once they are on course, the pigeons unhesitatingly cross lakes, valleys, and mountain ranges. Only if the topographic feature constitutes a really formidable obstacle, as in the case of some alpine massifs, are the pigeons diverted, and even then the effect is simply to make them fly back and forth along the face of the mountain until they gain sufficient altitude to cross. Wagner (1972) concludes that major topographical features usually operate to "make homing more difficult and not easier. There are no topographical structures or lines that lead pigeons home." These conclusions are, in general, consistent with those of Michener and Walcott.

Papi and Pardi (1968) and Papi et al. (1971b) have conducted test releases at sea. They report that the pigeons' orientation behavior was quite different from their usual behavior in releases on land. In particular, these authors think the birds often oriented in a direction perpendicular to the coast near their loft, even when the coast itself could not be seen. This would imply a strong aversion to flying over the sea. But Walcott (1972a and personal communication) has airplane-tracked some of his pigeons released 40 miles at sea and found nothing unusual about their orientational behavior; he saw no indication of a general orientation toward the nearest coast. Moreover, Walcott (1972b) has airplane-tracked a clock-shifted pigeon flying eastward from a release site in Connecticut that, when it came to the easternmost part of Cape Cod, headed on out to sea. Thus his results, like those of Wagner, suggest that once a bird is on course and flying steadily, even such major features of the landscape as the ocean shore often fail to divert it.

   d. Cage Experiments. None of the techniques discussed so far solve the problem of the difficulty in manipulating and controlling possible orientational cues when all experiments are performed with free-flying birds. Research on homing would benefit enormously if pigeons could be made to indicate their choice of the home direction in a circular cage. Unfortunately, numerous attempts by many investigators (e.g., most recently, Becker, 1968; Moreau and Pouyet, 1968) have failed. Thus it has seemed as though pigeons can readily learn to indicate compass bearings in a training cage when the sun is visible (Schmidt-Koenig, 1958, 1960), but that they resist learning to indicate the home direction. I remain unconvinced, however, and feel that a technique for accomplishing this task will be found. My associates and I have been experimenting with several cage designs that appear very promising; these do not require the birds to learn any special task—all that is necessary is that they attempt to escape from the cage.[22]

   Even if my optimism that pigeons can show us the home direction in a circular cage of the proper design is not borne out, another possibility should not be overlooked—that the pigeons might show a weak homeward tendency in some of the test situations already tried and discarded. Heretofore we have looked for orientation that is obviously nonrandom. But the experience of Merkel and Wiltschko with European robins should warn us to look more closely at supposedly random data. Their birds' jumps in circular cages are statistically random on at least 40% of the nights (Wiltschko, 1972), yet when all the nightly means are pooled for a second-order analysis, nonrandom orientation in a biologically meaningful direction is often found. Emlen (1970a) initially overlooked this method of analysis when he concluded that his indigo buntings gave random bearings under visually cueless conditions. Actually, if his data are reanalyzed in the manner of Merkel and Wiltschko, they show nonrandom orientation in the appropriate migratory direction (Fig. 29) (for a fuller discussion, see Emlen, in preparation). Perhaps a similar second-order reanalysis of "random" bearings of pigeons in circular cages will reveal homeward orientation.

   The work of Merkel and Wiltschko also points up the importance of rela-

---

[22] Chelazzi and Pardi (1972) have recently published a paper reporting homeward orientation by pigeons in a test cage.

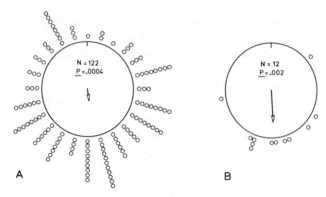

FIG. 29. (A) Pooled nightly means for indigo buntings in funnel cages in a "visually cueless" environment in autumn. Since these bearings are not all independent, and since some birds contributed more nightly means than others, the data are replotted (B) using only the mean of nightly means for each of the 12 birds; the significant southward orientation remains. (Plotted from data in Emlen, 1970a.)

tively small differences in cage design (Merkel, 1971). Though these investigators regularly get oriented behavior (on second-order analysis) in their cages in "visually cueless" situations, Perdeck (1963) and Wallraff (1966c) reported that they were unable to repeat the results. But the cages used by Merkel and Wiltschko record hops on radially arranged perches, whereas the cages of Perdeck and Wallraff had tangential perches. Recently Wallraff (1972b) has performed tests using a Merkel cage, and now he too gets oriented results. Experimenters with caged pigeons must be sensitive to the possibility that similarly small details of their designs may make the difference between success and failure.

VI. Concluding Remarks

The last few years have brought a new ferment to research on avian orientation. No longer is attention focused on only a few obvious environmental stimuli; indeed, a plethora of stimuli have been nominated for a role in the avian navigational system. No longer is it assumed that only a very few of these potential cues are really used by the birds; instead there is an openness to the possibility that there may be multiple components in the system. No longer is there a conviction that the components, whether few or many, must fit together in some single definitive way; rather there is a growing expectation that a variety of different hierarchies or weighting schemes are possible—that the components may be combined in alternative ways.

With these developments in our attitudes come many promising lines of attack for future research. I cannot presume to mention all of them here, but perhaps I may point to four categories that I personally find especially stimulating:

(1) The attempt to unravel the interrelationships between cues will be an intriguing challenge. How do the weighting schemes differ in different species? How do they differ in conspecifics of differing ages and experience? How does the weighting shift in response to different environmental conditions? Can birds be trained to use a particular weighting scheme under conditions in which it would not ordinarily be used? Do individuals of similar age and experience differ in the ways they respond to the various cues? What is the relative effectiveness of the various alternative systems? Does the availability of redundant cues, and their simultaneous use, improve accuracy? All these questions and others like them deserve our attention. They call for ingenuity in devising experiments that pit one cue against another, or that eliminate selected cues in carefully controlled contexts. They also call for rigorous attention to details of management of the experimental birds, so that unsuspected differences in motivation can be minimized. And they call for great perceptiveness in detecting small effects that may then be magnified in more precise experiments.

(2) At a time when a multiplicity of cues are being discussed, it is surely appropriate to look more closely at the sensory capabilities of the birds whose behavior we seek to understand. If there is a magnetic sense, how does it work and how sensitive is it? What are the capabilities of avian visual systems in contexts meaningful for homing—e.g., what are their powers of discrimination for objects several miles away? Does the recent evidence (Kreithen and Keeton, 1974b) that pigeons may be able to detect the e-vector of polarized light mean that they, like honeybees, can continue using the sun compass when the sun's disk is obscured by thick clouds, as long as some blue sky is visible? How responsive are flying birds to sounds on the ground?[23] Might birds be able to get orientational information from infrasound? Can birds learn to give orientational responses to smells, and if so, what are their powers of olfactory discrimination in this context? Can exposure to radar waves effect the initial orientation of pigeons, as Wagner (1972b) suggests? How do birds use their remarkable sensitivity to barometric pressure changes (Kreithen and Keeton, 1974a)? How precisely can birds detect changes in temperature, humidity, and the atmospheric electric field, and to what extent do they respond to them during homing or migratory orientation? The list of such questions could go on and on.

(3) The ontogeny of orientational and navigational behavior has so far

[23] For a discussion of the possible information content of such sounds, see D'Arms and Griffin (1972).

received much less attention than it deserves. Here is a potentially powerful avenue for examining the components of orientational systems, and for gaining understanding of the ways in which they come to be associated.

(4) Another neglected subject concerns the terminal portion of migratory flights. There is much debate whether migrants fly most of their route by simple compass orientation as many experiments suggest (e.g., experiments on hooded crows by Rüppell, 1944; Rüppell and Schuz, 1948; on immature starlings by Perdeck, 1958, 1967), or whether they are goal-orienting as Rabøl (1969, 1970, 1972) believes and as Perdeck's (1958, 1967) experiments on adult starlings suggest. But whatever the answer to that question may be, I am convinced that the documented high percentage of returns of individual migrant birds to the very same small areas where they resided in previous years can be explained only if we assume that near the end of their journey, if not before, migrants use goal-locating procedures analogous to (and perhaps identical with) those used in homing.

Times of ferment are always disturbing in some ways, and the present instance is no exception. It is, in a sense, unsettling that so many seemingly unconnected ideas are being debated in the absence of comprehensive hypotheses. But times of ferment are also invigorating, and no one can deny that there is currently a feeling of excitement among researchers in this field that promises new and more daring approaches to a problem that has intrigued and baffled mankind for centuries. With so much going on, I cannot begin to guess what a review such as this will say 5 years from now. But I look forward with relish to the revelations those 5 years will surely bring.

Acknowledgment

I thank the members of the Cornell Orientation Group, especially Irene Brown, Timothy Larkin, Andre Gobert, Stephen T. Emlen, Robin Alexander, Lindsey Goodloe, Donald Windsor, Melvin Kreithen, and Natalie Demong, for their help in gathering together appropriate data or in the preparation of this manuscript. I am also grateful to Monica Howland for preparing the drawings and to Bertha Blaker for typing the manuscript. My own research has been supported by Grants GB-13046X and GB-35199X from the National Science Foundation and by Hatch funds from the U.S. Department of Agriculture.

References

Able, K. P. 1974. Environmental influences on the orientation of free-flying nocturnal bird migrants. *Anim. Behav.* **22**, 224–238
Adler, H. E. 1963. Psychophysical limits of celestial navigation hypotheses. *Ergeb. Biol.* **26**, 235–252.
Aerts, J. 1969. "Pigeon Racing Advanced Techniques." Faber & Faber, London.
Alexander, J. R. 1970. The effect of directional training on initial orientation in pigeons. M.S. Thesis, Cornell Univ., Ithaca, New York.

Alexander, J. R. 1974. The effect of various phase-shifting experiments on homing in pigeons. Ph.D. Thesis, Cornell Univ., Ithaca, New York.

Alexander, J. R., and Keeton, W. T. 1972. The effect of directional training on initial orientation in pigeons. *Auk* **89**, 280–298.

Alexander, J., and Keeton, W. T. 1974. Clock-shifting effect on initial orientation of pigeons. *Auk* **91**, 370–374.

Bagg, A. M., Gunn, W. H., Miller, D. S., Nichols, J. T., Smith, W., and Wolfarth, F. P. 1950. Barometric pressure-patterns and spring bird migration. *Wilson Bull.* **62**, 5–19.

Baldaccini, N. E., Fiaschi, V., Fiore, L., and Papi, F. 1971. Initial orientation of directionally trained pigeons under overcast and sunny conditions. *Monit. Zool. Ital.* [*N.S.*] **5**, 53–63.

Bang, B. G., and Cobb, S. 1968. The size of the olfactory bulb in 108 species of birds. *Auk* **85**, 55–61.

Barker, W. E. 1958. "Pigeon Racing: A Practical Guide to the Sport," 3rd Ed. Racing Pigeon Publ. Co., London.

Barlow, J. S. 1964. Inertial navigation as a basis for animal navigation. *J. Theor. Biol.* **6**, 76–117.

Barlow, J. S. 1966. Inertial navigation in relation to animal navigation. *J. Inst. Navigation* **19**, 302–316.

Barlow, J. S. 1971. Remarks in a panel discussion: Unconventional theories of orientation. *Ann. N.Y. Acad. Sci.* **188**, 333–335.

Batschelet, E. 1965. "Statistical Methods for the Analysis of Problems in Animal Orientation and Certain Biological Rhythms." Amer. Inst. Biol. Sci., Washington, D.C.

Batschelet, E. 1972. Recent statistical methods for orientation data. *In* "Animal Orientation and Navigation." NASA SP-262, pp. 61–91. U.S. Gov. Printing Office, Washington, D.C.

Becker, G. 1963a. Ruheeinstellung nach der Himmelsrichtung eine Magnetfeldorientierung bei Termiten. *Naturwissenschaften* **50**, 455.

Becker, G. 1963b. Magnetfeld-Orientierung von Dipteren. *Naturwissenschaften* **50**, 664.

Becker, H. 1968. Weitere Versuche über Richtungstendenzen von Lachmöwen (*Larus ridibundus* L.) und Tauben nach Verfrachtung. *Z. Wiss. Zool.* **178**, 186–234.

Bellrose, F. C. 1958. Celestial orientation in wild mallards. *Bird Banding* **29**, 75–90.

Bellrose, F. C. 1963. Orientation behavior of four species of waterfowl. *Auk* **80**, 257–289.

Bellrose, F. C. 1967a. Radar in orientation research. *Proc. Int. Ornithol. Congr., 14th, Oxford, 1966* pp. 281–309.

Bellrose, F. C. 1967b. Orientation in waterfowl migration. *In* "Animal Orientation and Navigation" (R. M. Storm, ed.), pp. 73–99. Oregon State Univ. Press, Corvallis.

Bellrose, F. C. 1972. Possible steps in the evolutionary development of bird navigation. *In* "Animal Orientation and Navigation." NASA SP-262, pp. 223–257. U.S. Gov. Printing Office, Washington, D.C.

Bellrose, F. C., and Graber, R. R. 1963. A radar study of the flight directions of nocturnal migrants. *Proc. Int. Ornithol. Congr., 13th, Ithaca, N.Y.* pp. 362–389.

Bennett, M. F., and Huguenin, J. 1969. Geomagnetic effects on a circadian difference in reaction times in earthworms. *Z. Vergl. Physiol.* **63**, 440–445.

Benvenuti, S., Fiaschi, V., Fiore, L., and Papi, F. 1973a. Homing performance of inexperienced and directionally trained pigeons subjected to olfactory nerve section. *J. Comp. Physiol.* **83**, 81–92.

Benvenuti, S., Fiaschi, V., Fiore, L., and Papi, F. 1973b. Disturbances of homing behavior in pigeons experimentally induced by olfactory stimuli. *Monit. Zool. Ital. (N.S.)* **7**, 117–128.

Billings, S. M. 1968. Homing in Leach's Petrel. *Auk* **85**, 36–43.

Birner, M., Gernandt, D., Merkel, F. W., and Wiltschko, W. 1968. Verfrachtungsversuche mit einer Starenpopulation im Winter. *Natur Mus.* **98**, 507–514.

Blokpoel, H. 1970. A preliminary study on height and density of nocturnal fall migration. *Proc. World Conf. Bird Hazards Aircraft, Kingston, Ont., 1969* pp. 335–348.

Blough, D. S. 1956. Dark adaptation in the pigeon. *J. Comp. Physiol. Psychol.* **49**, 425–430.

Blough, P. M. 1971. The visual acuity of the pigeon for distant targets. *J. Exp. Anal. Behav.* **15**, 57–67.

Bochenski, Z., Dylewska, M., Gieszczykiewicz, J., and Sych, L. 1960. Homing experiments on birds. XI. Experiments with Swallows *Hirundo rustica* L. concerning the influence of earth magnetism and partial eclipse of the sun on their orientation. *Zesz. Nauk. W. J. Zool.* **5**, 125–130. (In Pol.; Engl. sum.)

Brown, F. A. 1971. Some orientational influences of nonvisual terrestrial electromagnetic fields. *Ann. N.Y. Acad. Sci.* **188**, 224–241.

Bruderer, B., and Steidinger, P. 1972. Methods of quantitative and qualitative analysis of bird migration with a tracking radar. *In* "Animal Orientation and Navigation." NASA SP-262, pp. 151–167. U.S. Gov. Printing Office, Washington, D.C.

Casamajor, J. 1926. Le mystérieux "sens de l'espace" chez les Pigeons voyageurs. *Nature (Paris)* No. 2748, 366–367.

Chalazonitis, N., Chagneux, R., and Arvanitaki, A. 1970. Rotation des segments externes des photorecepteurs dans le champ magnétique constant. *C. R. Acad. Sci., Ser. D* **271**, 130–133.

Chelazzi, G., and Pardi, L. 1972. Experiments on the homing behaviour of caged pigeons. *Monit. Zool. Ital. (N.S.)* **6**, 63–73.

Clark, C. L., Peck, R. A., and Hollander, W. F. 1948. Homing pigeon in electromagnetic fields. *J. Appl. Phys.* **19**, 1183.

Clarke, C. W. 1933. Night-flying homers of the signal corps. An experiment that resulted in a new race of homing pigeons. *Natur. Hist.* **33**, 409–418.

Daanje, A. 1936. Haben die Vögel einen Sinn für den Erdmagnetismus, wie Deklination, Inklination, und Intensität? *Ardea* **25**, 107–111.

Daanje, A. 1941. Heimfindeversuche und Erdmagnetismus. *Vogelzug* **12**, 15–17.

D'Arms, E., and Griffin, D. R. 1972. Balloonists' reports of sounds audible to migrating birds. *Auk* **89**, 269–279.

Davis, L. 1948. Remarks on: "The physical basis of bird navigation." *J. Appl. Phys.* **19**, 307–308.

de Vries, H. 1948. Die Reizschwelle der Sinnesorgane als physikalisches Problem. *Experientia* **4**, 205–213.

Dircksen, R. 1932. Die Biologie des Austernfischers, der Brandseeschwalbe und der Kustenseeschwalbe. *J. Ornithol* **80**, 427–521.

Downhower, J. F., and Windsor, D. M. 1971. Use of landmarks in orientation by Bank Swallows. *BioScience* **21** 570–572.

Drury, W. H., and Keith, J. A. 1962. Radar studies of songbird migration in coastal New England. *Ibis* **104**, 449–489.

Drury, W. H., and Nisbet, I. C. T. 1964. Radar studies of orientation of songbird migrants in southeastern New England. *Bird Banding* **35**, 69–119.

Eldarov, A. L., and Kholodov, Y. A. 1964. The effect of a constant magnetic field on the motor activity of birds. *Zh. Obschch. Biol.* **25**, 224–229. (In Russ.)

Emlen, S. T. 1967a. Migratory orientation of the Indigo Bunting, *Passerina cyanea.* Part I: Evidence for use of celestial cues. *Auk* **84**, 309–342.

Emlen, S. T. 1967b. Migratory orientation of the Indigo Bunting, *Passerina cyanea.* Part II: Mechanism of celestial orientation. *Auk* **84**, 463–489.

Emlen, S. T. 1967c. Orientation of Zugunruhe in the Rose-breasted Grosbeak, *Pheucticus ludovicianus.* *Condor* **69**, 203–205.

Emlen, S. T. 1969a. Bird migration: Influence of physiological state upon celestial orientation. *Science* **165**, 716–718.

Emlen, S. T. 1969b. The development of migratory orientation in young Indigo Buntings. *Living Bird* **8**, 113–126.

Emlen, S. T. 1970a. The influence of magnetic information on the orientation of the Indigo Bunting, *Passerina cyanea.* *Anim. Behav.* **18**, 215–224.

Emlen, S. T. 1970b. Celestial rotation: Its importance in the development of migratory orientation. *Science* **170**, 1198–1201.

Emlen, S. T. 1972. The ontogenetic development of orientation capabilities. *In* "Animal Orientation and Navigation." NASA SP-262, pp. 191–210. U.S. Gov. Printing Office, Washington, D.C.

Emlen, S. T. Migration: Orientation and navigation. *In* "Avian Biology" (D. S. Farner and J. R. King, eds.), Vol. 5. Academic Press, New York. In preparation.

Evans, P. R. 1966. Migration and orientation of passerine night migrants in northeast England. *J. Zool. (London)* **150**, 319–369.

Evans, P. R. 1972. Information on bird navigation obtained by British long-range radars. *In* "Animal Orientation and Navigation." NASA SP-262, pp. 139–149. U.S. Gov. Printing Office, Washington, D.C.

Exner, S. 1893. Negative versuchergebnisse über das Orientierungsvermögen der Brieftauben. *Sitzungsber. Kaiserl. Akad. Wiss. Wien,* **102**, 318–331. *Math.-Naturwiss. Kl., Abt. 1.*

Fromme, H. G. 1961. Untersuchungen über das Orientierungsvermögen nächtliche ziehender Kleinvögel (*Erithacus rubecula, Sylvia communis*). *Z. Tierpsychol.* **18**, 205–220.

Galifret, Y. 1968. Les diverse aires fonctionelles de la retine du Pigeon. *Z. Zellforsch. Mikrosk. Anat.* **86**, 535–545.

Gauthreaux, S. A. 1971. A radar and direct visual study of passerine spring migration in southern Louisiana. *Auk* **88**, 343–365.

Gauthreaux, S. A. 1972. Flight directions of passerine migrants in daylight and darkness: A radar and direct visual study. *In* "Animal Orientation and Navigation." NASA SP-262, pp. 129–137. U.S. Gov. Printing Office, Washington, D.C.

Gauthreaux, S. A., and Able, K. P. 1970. Wind and the direction of nocturnal songbird migration. *Nature (London)* **228**, 476–477.

Goodloe, L. 1974. Night homing in pigeons. Ph.D. Thesis, Cornell Univ., Ithaca, New York.

Gordon, D. A. 1948. Sensitivity of the homing pigeon to the magnetic field of the earth. *Science* **108**, 710–711.

Graue, L. C. 1963. The effect of phase shifts in the day-night cycle on pigeon homing at distances of less than one mile. *Ohio J. Sci.* **63**, 214–217.

Graue, L. C. 1965. Experience effect on initial orientation in pigeon homing. *Anim. Behav.* **13**, 149–153.

Graue, L. C. 1970. Orientation and distance in pigeon homing (*Columba livia*). *Anim. Behav.* **18**, 36–40.

Graue, L. C., and Pratt, J. G. 1959. Directional differences in pigeon homing in Sacramento, California and Cedar Rapids, Iowa. *Anim. Behav.* **7**, 201–208.

Griffin, D. R. 1940. Homing experiments with Leach's Petrels. *Auk* **57**, 61–74.

Griffin, D. R. 1943. Homing experiments with Herring Gulls and Common Terns. *Bird Banding* **14**, 7–33.

Griffin, D. R. 1952a. Bird navigation. *Biol. Rev. Cambridge Phil. Soc.* **27**, 359–400.

Griffin, D. R. 1952b. Airplane observations of homing pigeons. *Bull. Mus. Comp. Anat.* **107**, 411–440.

Griffin, D. R. 1955. Bird navigation. *In* "Recent Studies in Avian Biology" (A. Wolfson, ed.), pp. 154–197. Univ. of Illinois Press, Urbana.

Griffin, D. R. 1969. The physiology and geophysics of bird navigation. *Quart. Rev. Biol.* **4**, 255–276.

Griffin, D. R. 1972. Nocturnal bird migration in opaque clouds. *In* "Animal Orientation and Navigation." NASA SP-262, pp. 169–188. U.S. Gov. Printing Office, Washington, D.C.

Griffin, D. R. 1973. Oriented bird migration in or between opaque cloud layers. *Proc. Amer. Philos. Soc.* **117**, 117–141.

Griffin, D. R., and Goldsmith, T. H. 1955. Initial flight directions of homing birds. *Biol. Bull.* **108**, 264–276.

Griffin, D. R., and Hock, R. J. 1949. Airplane observations of homing birds. *Ecology* **30**, 176–198.

Gronau, J., and Schmidt-Koenig, K. 1970. Annual fluctuation in pigeon homing. *Nature (London)* **226**, 87–88.

Grubb, T. C. 1972. Smell and foraging in shearwaters and petrels. *Nature (London)* **237**, 404–405.

Gwinner, E. 1971. Orientierung. *In* "Grundriss der Vogelzugskunde" (E. Schüz *et al.*), pp. 299–348. Verlag Paul Parey, Berlin.

Hachet-Souplet, P. 1911. L'instinct du retour chez le pigeon voyageur. *Rev. Sci.* **29**, 231–238.

Haffner, M. E. 1966. Let's never sell the Trentons short. *Amer. Racing Pigeon News* **82**(3), 17–18.

Hamilton, W. J. 1962a. Celestial orientation in juvenile waterfowl. *Condor* **64**, 19–33.

Hamilton, W. J. 1962b. Bobolink migratory pathways and their experimental analysis under night skies. *Auk* **79**, 208–233.

Hamilton, W. J. 1966. Analysis of bird navigation experiments. *In* "Systems Analysis in Ecology" (K. E. F. Watt, ed.), pp. 147–178. Academic Press, New York.

Heinroth, O., and Heinroth, K. 1941. Das Heimfinde-Vermögen der Brieftauben. *J. Ornithol.* **89**, 213–256.

Henton, W. W. 1966. Suppression behavior to odorous stimuli in the pigeon. Ph.D. Thesis, Florida State Univ., Tallahassee.

Henton, W. W., Smith, J. C., and Tucker, D. 1966. Odor discrimination in pigeons. *Science* **153**, 1138–1139.

Hitchcock, H. B. 1952. Airplane observations of homing pigeons. *Proc. Amer. Phil. Soc.* **96**, 270–289.

Hitchcock, H. B. 1955. Homing flights and orientation in pigeons. *Auk* **72**, 355–373.

Hoffmann, K. 1953. Experimentelle Änderung des Richtungsfinden beim Star durch Beeinflussung der "innen Uhr." *Naturwissenschaften* **40**, 608–609.

Hoffmann, K. 1954. Versuche zu der in Richtungsfinden der Vögel enthaltenen Zeitschätzung. *Z. Tierpsychol.* **11**, 453–475.

Hoffmann, K. 1958. Repetition of an experiment on bird orientation. *Nature (London)* **181**, 1435–1437.

Hoffmann, K. 1959. Über den Einfluss verschiedener Faktoren auf die Heimkehrleistung von Brieftauben. *J. Ornithol.* **100**, 90–102.

Hoffmann, K. 1965. Clock mechanisms in celestial orientation of animals. *In* "Circadian Clocks" (J. Aschoff, ed.), pp. 426–441. North-Holland Publ., Amsterdam.

Hoffmann, K. 1971. Biological clocks in animal orientation and in other functions. *Proc. Int. Symp. Circadian Rhythmicity, Wageningen* pp. 175–205.

Hong, F. T., Mauzerall, D., and Mauro, A. 1971. Magnetic anisotropy and the orientation of retinal rods in a homogeneous magnetic field. *Proc. Nat. Acad. Sci. U.S.* **68**, 1283–1285.

Huizinger, E. 1935. Durchschneidung aller Bogengänge bei der Taube. *Pfluegers Arch. Gesamte Physiol. Menschen Tiere* **236**, 52–58.

Hutton, A. N. 1964. "Pigeon Racing," 3rd Ed. Lockwood, London.

Hutton, A. N. 1966. "Pigeon Lore," 2nd Ed. Lockwood, London.

Ising, G. 1945. Die physikalische Möglichkeit eines Tierischen Orientierungsinnes auf Basis der Erdrotation. *Ark. Mat. Astron. Fys.* **32**, 1–23.

Keeton, W. T. 1969. Orientation by pigeons: Is the sun necessary? *Science* **165**, 922–928.

Keeton, W. T. 1970a. Orientation by pigeons. *Science* **168**, 153.

Keeton, W. T. 1970b. Do pigeons determine latitudinal displacement from the sun's altitude? *Nature (London)* **227**, 626–627.

Keeton, W. T. 1970c. "Distance effect" in pigeon orientation: An evaluation. *Biol. Bull.* **139**, 510–519.

Keeton, W. T. 1970d. Comparative orientational and homing performance of single pigeons and small flocks. *Auk* **87**, 797–799.

Keeton, W. T. 1971a. Magnets interfere with pigeon homing. *Proc. Nat. Acad. Sci. U.S.* **68**, 102–106.

Keeton, W. T. 1971b. Remarks in a panel discussion: Unconventional theories of orientation. *Ann. N.Y. Acad. Sci.* **188**, 331–333, 338–340.

Keeton, W. T. 1972. Effects of magnets on pigeon homing. *In* "Animal Orientation and Navigation." NASA SP-262, pp. 579–594. U.S. Gov. Printing Office, Washington, D.C.

Keeton, W. T. 1973. Release-site bias as a possible guide to the "map" component in pigeon homing. *J. Comp. Physiol.* **86**, 1–16.

Keeton, W. T., and Gobert, A. 1970. Orientation by untrained pigeons requires the sun. *Proc. Nat. Acad. Sci. U.S.* **65**, 853–856.

Kenyon, K. W., and Rice, D. W. 1958. Homing of Laysan Albatrosses. *Condor* **60**, 3–6.

Kluijver, H. W. 1935. Ergebnisse eines Versuches über das Heimfindevermögen von Staren. *Ardea* **24**, 227–239.

Kramer, G. 1949. Über Richtungstendenzen bei der nächtlichen Zugunruhe gekäfigten Vögel. *In* "Ornithologie als biologische Wissenschaft" (E. Mayr and E. Schüz, eds.), pp. 269–283. Winter, Heidelberg.

Kramer, G. 1950a. Orientierte Zugaktivität gekäfigter Singvögel. *Naturwissenschaften* **37**, 188.

Kramer, G. 1950b. Weitere Analyse der Faktoren, welche die Zugaktivität des gekäfigten Vogels orientieren. *Naturwissenschaften* **37**, 377–378.

Kramer, G. 1951. Eine neue Methode zur Erforschung der Zugorientierung und die bisher damit erzielten Ergebnisse. *Proc. Int. Ornithol. Congr., 10th, Uppsala* pp. 271–280.

Kramer, G. 1952. Experiments on bird orientation. *Ibis* **94**, 265–285.

Kramer, G. 1953a. Wird die Sonnenhöhe bei der Heimfindeorientierung verwertet? *J. Ornithol.* **94**, 201–219.

Kramer, G. 1953b. Die Sonnenorientierung der Vögel. *Verh. Deut. Zool. Ges. Freiburg, 1952* pp. 72–84.

Kramer, G. 1954. Einfluss von Temperatur und Erfahrung auf das Heimfindevermögen von Brieftauben. *J. Ornithol.* **95**, 343–347.

Kramer, G. 1955. Ein weiterer Versuch, die Orientierung von Brieftauben durch jahreszeitliche Änderung der Sonnenhöhe zu beeinflussen. Gleichzeitig eine Kritik der Theorie des Versuchs. *J. Ornithol.* **96**, 173–185.

Kramer, G. 1957. Experiments in bird orientation and their interpretation. *Ibis* **99**, 196–227.

Kramer, G. 1959. Recent experiments on bird orientation. *Ibis* **101**, 399–416.

Kramer, G., and von Saint Paul, U. 1950. Ein Wesentlicher Bestandteil der Orientierung der Reisetaube: die Richtungsdressur. *Z. Tierpsychol.* **7**, 620–631.

Kramer, G., and von St. Paul, U. 1954. Das Heimkehrvermögen gekäfigter Brieftauben. *Ornithol. Beobachter* **51**, 3–12.

Kramer, G., and von St. Paul, U. 1956. Weitere Erfahrungen über den "Wintereffekt" beim Heimfindevermögen von Brieftauben. *J. Ornithol.* **97**, 353–370.

Kramer, G., Pratt, J. G., and von St. Paul, U. 1957. Two-direction experiments with homing pigeons and their bearing on the problem of goal orientation. *Amer. Natur.* **91**, 37–48.

Kramer, G., Pratt, J. G., and von St. Paul, U. 1958. Neue Untersuchungen über den 'Richtungseffekt.' *J. Ornithol.* **99**, 178–191.

Kramer, G., von St. Paul, U., and Wallraff, H. G. 1959. Über die Heimfindeleistung von unter Sichtbegrenzung aufgewachsenen Brieftauben. *Verh. Deut. Zool. Ges. Frankfurt, 1958* pp. 168–176.

Kreithen, M. L., and Keeton, W. T. 1974a. Detection of changes in atmospheric pressure by the homing pigeon, *Columba livia. J. Comp. Physiol.* **89**, 73–82.

Kreithen, M. L., and Keeton, W. T. 1974b. Detection of polarized light by the homing pigeon, *Columba livia. J. Comp. Physiol.* **89**, 83–92.

Kreithen, M. L., and Keeton, W. T. (1974c). Attempts to condition homing pigeons to magnetic stimuli. *J. Comp. Physiol.* In press.

Lack, D. 1960. The influence of weather on passerine migration: A review. *Auk* **77**, 171–209.

Lack, D. 1969. Drift migration: A correction. *Ibis* **111**, 253–255.

Lack, D., and Lockley, R. M. 1938. Skokholm Bird Observatory homing experiments. I. 1936–1937. Puffins, Storm Petrels, and Manx Shearwaters. *Brit. Birds* **31**, 242–248.

Levi, W. M. 1963. "The Pigeon," 1963 Ed. Levi Publ. Co., Sumter, South Carolina.

Lindauer, M., and Martin, H. 1968. Die Schwereorientierung der Bienen unter dem Einfluss des Erdmagnetfeldes. *Z. Vergl. Physiol.* **60**, 219–243.

Lindauer, M., and Martin, H. 1972. Magnetic effect on dancing bees. *In* "Animal Orientation and Navigation." NASA SP-262, pp. 559–567. U.S. Gov. Printing Office, Washington, D.C.

McDonald, D. L. 1968. Bird orientation: A method of study. *Science* **161**, 486–487.

McDonald, D. L. 1972. Some aspects of the use of visual cues in directional training of homing pigeons. *In* "Animal Orientation and Navigation." NASA SP-262, pp. 293–304. U.S. Gov. Printing Office, Washington, D.C.

McDonald, D. L. 1973. The role of shadows in directional training and homing of pigeons, *Columba livia. J. Exp. Zool.* **183**, 267–280.

Matthews, G. V. T. 1951a. The sensory basis of bird navigation. *J. Inst. Navigation* **4**, 260–275.

Matthews, G. V. T. 1951b. The experimental investigation of navigation in homing pigeons. *J. Exp. Biol.* **28**, 508–536.

Matthews, G. V. T. 1952. An investigation of homing ability in two species of gulls. *Ibis* **94**, 243–264.

Matthews, G. V. T. 1953a. Sun navigation in homing pigeons. *J. Exp. Biol.* **30**, 243–267.

Matthews, G. V. T. 1953b. The orientation of untrained pigeons: A dichotomy in the homing process. *J. Exp. Biol.* **30**, 268–276.

Matthews, G. V. T. 1953c. Navigation in the Manx Shearwater. *J. Exp. Biol.* **30**, 370–396.

Matthews, G. V. T. 1955a. "Bird Navigation." Cambridge Univ. Press, London and New York.

Matthews, G. V. T. 1955b. An investigation of the "chronometer" factor in bird navigation. *J. Exp. Biol.* **32**, 39–58.

Matthews, G. V. T. 1961. "Nonsense" orientation in Mallard, *Anas platyrynchos,* and its relation to experiments on bird navigation. *Ibis* **103a**, 211–230.

Matthews, G. V. T. 1963a. The astronomical bases of "nonsense" orientation. *Proc. Int. Ornithol. Congr., 13th, Ithaca, N.Y.* pp. 415–429.

Matthews, G. V. T. 1963b. The orientation of pigeons as affected by the learning of landmarks and by the distance of displacement. *Anim. Behav.* **11**, 310–317.

Matthews, G. V. T. 1964. Individual experience as a factor in the navigation of Manx Shearwaters. *Auk* **81**, 132–146.

Matthews, G. V. T. 1968. "Bird Navigation," 2nd Ed. Cambridge Univ. Press, London and New York.

Matthews, G. V. T. 1970. Response to: Do pigeons determine latitudinal displacement from the sun's altitude? *Nature (London)* **227**, 627.

Matthews, G. V. T., Eygenraam, J. A., and Hoffmann, L. 1963. Initial direction tendencies in the European Greenwinged Teal. *Wildfowl Trust Annu. Rep.* **14**, 120–123.

Merkel, F. W. 1971. Orientation behavior of birds in Kramer cages under different physical cues. *Ann. N.Y. Acad. Sci.* **188**, 283–294.

Merkel, F. W., and Fromme, H. G. 1958. Untersuchungen über das Orientierungsvermögen nächtlich ziehender Rotkehlchen (*Erithacus rubecula*). *Naturwissenschaften* **45**, 499–500.

Merkel, F. W., and Wiltschko, W. 1965. Magnetismus und Richtungsfinden zugunruhiger Rotkehlchen (*Erithacus rubecula*). *Vogelwarte* **23**, 71–77.

Merkel, F. W., Fromme, H. G., and Wiltschko, W. 1964. Nichtvisuelles Orientierungsvermögen bei nächtlich zugunruhigen Rotkehlchen. *Vogelwarte* **22**, 168–173.

Mewaldt, L. R., and Rose, R. G. 1960. Orientation of migratory restlessness in the White-crowned Sparrow. *Science* **131**, 105–106.

Mewaldt, L. R., Morton, M. L., and Brown, I. L. 1964. Orientation of migratory restlessness in *Zonotrichia. Condor* **66**, 377–417.

Meyer, M. E. 1964. Discriminative basis for astronavigation in birds. *J. Comp. Physiol. Psychol.* **58**, 403–406.

Meyer, M. E. 1966. The internal clock hypothesis for astronavigation in homing pigeons. *Psychon. Sci.* 5, 259–260.

Meyer, M. E., and Lambe, D. R. 1966. Sensitivity of the pigeon to changes in the magnetic field. *Psychon. Sci.* 5, 349–350.

Michener, M. C., and Walcott, C. 1966. Navigation of single homing pigeons: Airplane observations by radio tracking. *Science* 154, 410–413.

Michener, M. C., and Walcott, C. 1967a. Homing of single pigeons—analysis of tracks. *J. Exp. Biol.* 47, 99–131.

Michener, M. C., and Walcott, C. 1967b. Homing in pigeons. *Science* 155, 1136.

Miselis, R., and Walcott, C. 1970. Locomotor activity rhythms in homing pigeons (*Columba livia*). *Anim. Behav.* 18, 544–561.

Moreau, G., and Pouyet, J.-C. 1968. Possibilité de mise en évidence du rôle du soleil dans l'orientation des pigeons voyageurs captifs. *Alauda* 36, 108–120.

Murray, B. 1967. Homing in pigeons. *Science* 155, 1135–1136.

Neville, J. R. 1955. An experimental study of magnetic factors possibly concerned with bird navigation. Ph.D. Thesis, Stanford Univ., Palo Alto, California.

Nisbet, I. C. T. 1955. Atmospheric turbulence and bird flight. *Brit. Birds* 48, 557–559.

Nisbet, I. C. T., and Drury, W. H. 1967. Orientation of spring migrants studied by radar. *Bird Banding* 38, 173–186.

Nisbet, I. C. T., and Drury, W. H. 1968. Short-term effects of weather on bird migration: A field study using multivariate statistics. *Anim. Behav.* 16, 496–530.

Orgel, A. R., and Smith, J. C. 1954. Test of the magnetic theory of homing. *Science* 120, 891–892.

Papi, F., and Pardi, L. 1968. Are pigeons able to home when released over the sea? *Monit. Zool. Ital.* [*N.S.*] 2, 217–231.

Papi, F., Fiore, L., Fiaschi, V., and Benvenuti, S. 1971a. The influence of olfactory nerve section on the homing capacity of carrier pigeons. *Monit. Zool. Ital.* [*N.S.*] 5, 265–267.

Papi, F., Fiore, L., Fiaschi, V., and Baldaccini, N. E. 1971b. Orientation of pigeons released over the sea. *Z. Vergl. Physiol.* 73, 317–338.

Papi, F., Fiore, L., Fiaschi, V., and Benvenuti, S. 1972. Olfaction and homing in pigeons. *Monit. Zool. Ital.* [*N.S.*] 6, 85–95.

Papi, F., Fiore, L., Fiaschi, V., and Benvenuti, S. 1973a. An experiment for testing the hypothesis of olfactory navigation of homing pigeons. *J. Comp. Physiol.* 83, 93–102.

Papi, F., Fiaschi, V., Benvenuti, S., and Baldaccini, N. E. 1973b. Pigeon homing: Outward journey detours influence the initial orientation. *Monit. Zool. Ital.* (*N.S.*) 7, 129–133.

Pennycuick, C. J. 1960. The physical basis of astronavigation in birds: Theoretical considerations. *J. Exp. Biol.* 37, 573–593.

Pennycuick, C. J. 1961. Sun navigation in birds? *Nature (London)* 190, 1026.

Perdeck, A. C. 1958. Two types of orientation in migrating Starlings *Sturnus vulgaris* L. and Chaffinches *Fringilla coelebs* L., as revealed by displacement experiments. *Ardea* 46, 1–37.

Perdeck, A. C. 1963. Does navigation without visual clues exist in Robins? *Ardea* 51, 91–104.

Perdeck, A. C. 1967. Orientation of Starlings after displacement to Spain. *Ardea* 55, 194–202.

Petit, L., and Depauw, A. 1952. "The Practical Side of Pigeon Racing." Depauw, Denderleeuw, Belgium.

Picton, H. D. 1966. Some responses of *Drosophila* to weak magnetic and electrostatic fields. *Nature (London)* 211, 303–304.

Pratt, J. G., and Thouless, R. H. 1955. Homing orientation in pigeons in relation to opportunity to observe the sun before release. *J. Exp. Biol.* 32, 140–157.

Pratt, J. G., and Wallraff, H. G. 1958. Zwei-Richtungs-Versuche mit Brieftauben Langstreckenflüge auf der Nord-Süd-achse in Westdeutschland. *Z. Tierpsychol.* 15, 332–339.

Rabøl, J. 1969. Orientation of autumn migrating garden warblers (*Sylvia borin*) after displacement from western Denmark (*Blåvand*) to eastern Sweden (*Ottenby*). A preliminary experiment. *Dan. Ornithol. Foren. Tidsskr.* 63, 93–104.

Rabøl, J. 1970. Displacement and phaseshift experiments with night-migrating passerines. *Ornis Scand.* 1, 27–43.

Rabøl, J. 1972. Displacement experiments with night-migrating passerines. *Z. Tierpsychol.* 30, 14–25.

Rawson, K. S., and Rawson, A. M. 1955. The orientation of homing pigeons in relation to change in sun declination. *J. Ornithol.* 96, 168–172.

Reille, A. 1968. Essai de mise en évidence d'une sensibilité du pigeon au champ magnétique à l'aide d'un conditionnement nociceptif. *J. Physiol. (Paris)* 60, 85–92.

Richardson, W. J. 1971. Spring migration and weather in eastern Canada: A radar study. *Amer. Birds* 25, 684–690.

Richardson, W. J. 1972. Autumn migration and weather in eastern Canada: A radar study. *Amer. Birds* 26, 10–17.

Riviere, B. B. 1929. The "homing faculty" in pigeons. *Verh. Ornithol. Kongr., 6th,* Copenhagen pp. 535–555.

Rüppell, W. 1934. Heimfinde-Versuche mit Rauchschwalben (*Hirundo rustica*) und Mehlschwalben (*Delichon urbica*) von H. Warnat (Berlin-Charlottenburg). *Vogelzug* 5, 161–166.

Rüppell, W. 1935. Heimfindeversuche mit Staren 1934. *J. Ornithol.* 83, 462–524.

Rüppell, W. 1936. Heimfindeversuche mit Staren und Schwalben 1935. *J. Ornithol.* 84, 180–198.

Rüppell, W. 1937. Heimfindeversuche mit Staren, Rauchschwalben, Wendhälsen, Rotruckwürgen und Habichten 1936. *J. Ornithol.* 85, 120–135.

Rüppell, W. 1944. Versuche über Heimfinden ziehender Nebelkrähen nach Verfrachtung. *J. Ornithol.* 92, 106–133.

Rüppell, W., and Shüz, E. 1948. Ergebnis der Verfrachtung von Nebelkrähen (*Corvus corone cornix*) wahrend des Wegzuges. *Vogelwarte* 15, 30–36.

Sauer, E. G. F. 1957. Die Sternenorientierung nächtlich ziehender Grasmücken (*Sylvia atricapilla, borin* und *curruca*). *Z. Tierpsychol.* 14, 29–70.

Sauer, E. G. F. 1961. Further studies on the stellar orientation of nocturnally migrating birds. *Psychol. Forsch.* 26, 224–244.

Sauer, E. G. F. 1963. Migration habits of Golden Plovers. *Proc. Int. Ornithol. Congr., 13th, Ithaca, N.Y.* pp. 454–467.

Sauer, E. G. F., and Sauer, E. M. 1959. Nächtliche Zugorientierung europäischer Vögel in Südwestafrika. *Vogelwarte* 20, 4–31.

Sauer, E. G. F., and Sauer, E. M. 1960. Star navigation of nocturnal migrating birds. The 1958 planetarium experiments. *Cold Spring Harbor Symp. Quant. Biol.* 25, 463–473.

Schifferli, A. 1942. Verfrachtungversuche mit Alpenseglern (*Micropus m. melba*) Solothurn-Lissabon. *Ornithol. Beobachter* 39, 145–150.

Schifferli, A. 1951. Transportversuche mit Alpenseglern (*Apus melba*) nach Nairobi. *Ornithol. Beobachter* **48**, 183–184.

Schlichte, H. J. 1973. Untersuchungen über die Bedeutung optischer Parameter für das Heimkehrverhalten der Brieftaube. *Z. Tierpsychol.* **32**, 257–280.

Schlichte, H. J., and Schmidt-Koenig, K. 1971. Zum Heimfindevermögen der Brieftaube bei erschwerter optischer Wahrnehmung. *Naturwissenschaften* **58**, 329–330.

Schmidt-Koenig, K. 1958. Experimentelle Einflussnahme auf die 24-Stunden-Periodik bei Brieftauben und deren Auswirkungen unter besonderer Berücksichtigung des Heimfindevermögens. *Z. Tierpsychol.* **15**, 301–331.

Schmidt-Koenig, K. 1960. Internal clocks and homing. *Cold Spring Harbor Symp. Quant. Biol.* **25**, 389–393.

Schmidt-Koenig, K. 1961. Die Sonne als Kompass im Heim-Orientierungssystem der Brieftauben. *Z. Tierpsychol.* **18**, 221–244.

Schmidt-Koenig, K. 1963a. On the role of the loft, the distance and site of release in pigeon homing (The "cross loft experiment"). *Biol. Bull.* **125**, 154–164.

Schmidt-Koenig, K. 1963b. Neuere Aspekte über die Orientierungsleistungen von Brieftauben. *Ergeb. Biol.* **26**, 286–297.

Schmidt-Koenig, K. 1964. Initial orientation and distance of displacement in pigeon homing. *Nature (London)* **201**, 638.

Schmidt-Koenig, K. 1965. Current problems in bird orientation. In "Advances in the Study of Behavior" (D. S. Lehrman, R. A. Hinde, and E. Shaw, eds.), Vol. 1, pp. 217–278. Academic Press, New York.

Schmidt-Koenig, K. 1966. Über die Entfernung als Parameter bei der Anfangsorientierung der Brieftaube. *Z. Vergl. Physiol.* **52**, 33–55.

Schmidt-Koenig, K. 1968. Entfernung und Genauigkeit der Anfangsorientierung von Brieftauben. *Z. Vergl. Physiol.* **58**, 344–346.

Schmidt-Koenig, K. 1970. Entfernung und Heimkehrverhalten der Brieftaube. *Z. Vergl. Physiol.* **68**, 39–48.

Schmidt-Koenig, K. 1971. Remarks in a panel discussion: Unconventional theories of orientation. *Ann. N.Y. Acad. Sci.* **188**, 338–339.

Schmidt-Koenig, K. 1972. New experiments on the effect of clock shifts on homing in pigeons. In "Animal Orientation and Navigation." NASA SP-262, pp. 275–282. U.S. Gov. Printing Office, Washington, D.C.

Schmidt-Koenig, K., and Schlichte, H. J. 1972. Homing in pigeons with reduced vision. *Proc. Nat. Acad. Sci. U.S.* **69**, 2446–2447.

Schmidt-Koenig, K., and Walcott, C. 1973. Flugwege und Verbleib von Brieftauben mit getrübten Haftschalen. *Naturwissenschaften* **60**, 108–109.

Schneider, F. 1963. Ultroptische Orientierung des Maifäfers (*Melolntha vulgaris* F.) in Künstlichen elektrischen und magnetischen Feldern. *Ergeb. Biol.* **40**, 252–279.

Schreiber, B., Gualtierotti, T., and Mainardi, D. 1962. Some problems of cerebellar physiology in migratory and sedentary birds. *Anim. Behav.* **10**, 42–47.

Schüz, E., Berthold, P., Gwinner, E., and Oelke, H. 1971. "Grundriss der Vogelzugskunde." Verlag Paul Parey, Berlin.

Shumake, S. A., Smith, J. C., and Tucker, D. 1969. Olfactory intensity-difference thresholds in the pigeon. *J. Comp. Physiol. Psychol.* **67**, 64–69.

Shumakov, M. E. 1965. Preliminary results of the investigation of migrational orientation of passerine birds by the round-cage method. In "Bionika," pp. 371–378. Moscow. (In Russ.)

Shumakov, M. E. 1967. Experiments to determine the possibility of magnetic orientation in birds. In "Questions on *Bionics*," pp. 519–523. Vopr. Moscow (In Russ.)

Slepian, J. 1948. Physical basis of bird navigation. *J. Appl. Phys.* **19**, 306.

Smith, J. C., and Tucker, D. 1969. Olfactory mediation of immediate X-ray detection. *In* "Olfaction and Taste 3" (C. Pfaffman, ed.), pp. 288–298. Rockefeller Univ. Press, New York.

Snyder, L. R. G. 1970. A criticism of sun navigation hypotheses from the standpoint of circadian rhythm research. Senior Honors Thesis, Harvard Univ., Cambridge, Massachusetts.

Snyder, T. H. 1971. The Haffner 1000 m. strain. *Amer. Racing Pigeon News* **87**(3), 18–21.

Sobol, E. D. 1930. Orienting ability of carrier pigeons with injured labyrinths. *Mil.-Med. Zh. USSR* **1**, 75. [*Biol. Abstr.* **8**, 15425.]

Sonnberg, A., and Schmidt-Koenig, K. 1970. Zuer Auslese qualifizierter Brieftauben durch Übungsflüge. *Z. Tierpsychol.* **27**, 622–625.

Southern, W. E. 1959. Homing of Purple Martins. *Wilson Bull.* **71**, 254–261.

Southern, W. E. 1968. Experiments on the homing ability of Purple Martins. *Living Bird* **7**, 71–84.

Southern, W. E. 1969a. Orientation behavior of Ring-billed Gull chicks and fledglings. *Condor* **71**, 418–425.

Southern, W. E. 1969b. Gull orientation behavior: Influence of experience, sex, age, and group releases. *Jack-Pine Warbler* **47**, 34–43.

Southern, W. E. 1969c. Sky conditions in relation to Ring-billed and Herring Gull orientation. *Trans. Ill. Acad. Sci.* **62**, 342–349.

Southern, W. E. 1971. Gull orientation by magnetic cues: A hypothesis revisited. *Ann. N.Y. Acad. Sci.* **188**, 295–311.

Southern, W. E. 1972a. Influence of disturbances in the earth's magnetic field on Ring-billed Gull orientation. *Condor* **74**, 102–105.

Southern, W. E. 1972b. Magnets disrupt the orientation of juvenile Ring-billed Gulls. *BioScience* **22**, 476–479.

Spaepen, J., and Dachy, P. 1952. Le problème de l'orientation chez les oiseaux migrateurs. II. Expériences préliminaires effectuées sur des Martinets noirs, *Apus apus* L. *Gerfaut* **42**, 54–59.

Spaepen, J., and Dachy, P. 1953. Het Oriëntatieprobleem bij de Trekvogels. III. Verdere homingproeven met Gierzwaluwen (*Apus apus* L.). *Gerfaut* **43**, 327–332.

Spaepen, J., and Fragniere, H. 1952. Le problème de l'orientation chez les oiseaux migrateurs. I. Expériences préliminaires effectuées sur des Martinets alpins, *Apus melba* (L.). *Gerfaut* **42**, 49–54.

Stager, K. E. 1964. The role of olfaction in food location by the Turkey Vulture (*Cathartes aura*). *Los Angeles County Mus. Contrib. Sci.* No. 81, 63 pp.

Steidinger, P. 1968. Radarbeobachtungen über die Richtung und deren Streuung beim nächtlichen Vogelzug im Schweizerischen Mittelland. *Ornithol. Beobachter* **65**, 197–226.

Steidinger, P. 1972. Der Einfluss des Windes auf die Richtung des nächtlichen Vogelzuges. *Ornithol. Beobachter* **69**, 20–39.

Stresemann, E. 1935. Haben die Vögel Ortssinn? *Ardea* **24**, 213–226.

Suter, R. D., and Rawson, K. S. 1968. Circadian activity rhythm in the deer mouse, *Peromyscus:* Effect of deuterium oxide. *Science* **160**, 1011–1014.

Talkington, L. 1964. On bird navigation. *Amer. Ass. Advan. Sci. Meet., Montreal.* (Mimeo.)

Tansley, K. 1965. "Vision in Vertebrates." Cox & Wyman, London.

Thauzies, A. 1898. L'orientation. *Rev. Sci.* [*4th Ser.*] **9**, 392–397.

Thorpe, W. H. 1949. Recent biological evidence for the methods of bird orientation. *Proc. Linn. Soc. Lond.* **160**, 85–94.

Tucker, D. 1965. Electrophysiological evidence for olfactory function in birds. *Nature* (*London*) **207**, 34–36.

U.S. Army 1923. "The Pigeoneer." Training Manual No. 32. U.S. Gov. Printing Office, Washington, D.C.

U.S. War Department 1945. "The Homing Pigeon." Techn. Manual No. 11–410. U.S. Gov. Printing Office, Washington, D.C.

van Riper, W., and Kalmbach, E. R. 1952. Homing not hindered by wing magnets. *Science* **115**, 577–578.

Varian, R. H. 1948. Remarks on: "A preliminary study of a physical basis of bird navigation." *J. Appl. Phys.* **19**, 306–307.

Viguier, C. 1882. Le sens d'orientation et ses organes chez les animaux et chez l'homme. *Rev. Phil.* **14**, 1–36.

Violette, 1958. "Practical Course on the General Management of Pigeons." Racing Pigeon Publ. Co., London.

Vleugel, D. A. 1954. Waarnemingen over de nachtrek van lijsters (*Turdus*) en hun waarschijnlijke oriëntering. *Limosa* **27**, 1–19.

Vleugel, D. A. 1959. Über die wahrscheinlichste Methode der Wind-Orientierung ziehender Buchfinken (*Fringilla coelebs*). *Ornithol. Fenn.* **36**, 78–88.

Vleugel, D. A. 1962. Über nächtlichen Zug von Drosseln und ihre Orientierung. *Vogelwarte* **21**, 307–313.

von Saint Paul, U. 1962. Das Nachtfliegen von Brieftauben. *J. Ornithol.* **103**, 337–343.

Wagner, G. 1968. Topographisch bedingte zweigipflige und schiefe Kreisverteilungen bei der Anfangsorientierung verfrachteter Brieftauben. *Rev. Suisse Zool.* **75**, 682–690.

Wagner, G. 1970. Verfolgung von Brieftauben im Helikopter. *Rev. Suisse Zool.* **77**, 39–60.

Wagner, G. 1972a. Topography and pigeon orientation. *In* "Animal Orientation and Navigation." NASA SP-262, pp. 259–273. U.S. Gov. Printing Office, Washington, D.C.

Wagner, G. 1972b. Untersuchungen über das Orientierungsverhalten von Brieftauben unter Radar-Bestrahlung. *Rev. Suisse Zool.* **79**, 229–244.

Walcott, C. 1972a. The navigation of homing pigeons: Do they use sun navigation? *In* "Animal Orientation and Navigation." NASA SP-262, pp. 283–292. U.S. Gov. Printing Office, Washington, D.C.

Walcott, C. 1972b. Bird navigation. *Natur. Hist.* **81**(6), 32–43.

Walcott, C., and Green, R. P. 1974. Orientation of homing pigeons altered by a change in the direction of an applied magnetic field. *Science* **184**, 180–182.

Walcott, C., and Michener, M. C. 1967. Analysis of tracks of single homing pigeons. *Proc. Int. Ornithol. Congr., 14th, Oxford* pp. 311–329.

Walcott, C., and Michener, M. C. 1971. Sun navigation in homing pigeons—attempts to shift sun coordinates. *J. Exp. Biol.* **54**, 291–316.

Walcott, C., and Schmidt-Koenig, K. 1973. The effect of anesthesia during displacement on the homing performance of pigeons. *Auk* **90**, 281–286.

Wallraff, H. G. 1959a. Örtlich und zeitlich bedingte Variabilität des Heimkehrverhaltens von Brieftauben. *Z. Tierpsychol.* **16**, 513–544.

Wallraff, H. G. 1959b. Über den Einfluss der Erfahrung auf das Heimfindevermögen von Brieftauben. *Z. Tierpsychol.* **16**, 424–444.

Wallraff, H. G. 1960a. Über Zusammenhänge des Heimkehrverhaltens von Brieftauben mit meteorologischen und geophysikalischen Faktoren. *Z. Tierpsychol.* 17, 82–113.

Wallraff, H. G. 1960b. Können Grasmücken mit Hilfe des Sternenhimmels navigieren? *Z. Tierpsychol.* **17**, 165–177.

Wallraff, H. G. 1960c. Does celestial navigation exist in animals? *Cold Spring Harbor Symp. Quant. Biol.* **25**, 451–461.

Wallraff, H. G. 1965. Über das Heimfindevermögen von Brieftauben mit durchtrennten Bogengängen. *Z. Vergl. Physiol.* **50**, 313–330.

Wallraff, H. G. 1966a. Über die Anfangsorientierung von Brieftauben unter geschlossener Wolkendecke. *J. Ornithol.* **107**, 326–336.

Wallraff, H. G. 1966b. Über die Heimfindeleistungen von Brieftauben nach Haltung in verschiedenartig abgeschirmten Volieren. *Z. Vergl. Physiol.* **52**, 215–259.

Wallraff, H. G. 1966c. Versuche zur Frage der gerichteten Nachtzug-Aktivität von gekäfigten Singvögeln. *Verh. Deut. Zool. Ges. Jena, 1965* pp. 338–355.

Wallraff, H. G. 1967. The present status of our knowledge about pigeon homing. *Proc. Int. Ornithol. Congr., 14th, Oxford* pp. 331–358.

Wallraff, H. G. 1969. Über das Orientierungsvermögen von Vögeln unter natürlichen und künstlichen Sternenmustern. Dressurversuche mit Stockenten. *Verh. Deut. Zool. Ges. Innsbruck, 1968* pp. 348–357.

Wallraff, H. G. 1970a. Weitere Volierenversuche mit Brieftauben: Wahrscheinlicher Einfluss dynamischer Faktoren der Atmosphäre auf die Orientierung. *Z. Vergl. Physiol.* **68**, 182–201.

Wallraff, H. G. 1970b. Über die Flugrichtungen verfrachteter Brieftauben in Abhängigkeit vom Heimatort und vom Ort der Freilassung. *Z. Tierpsychol.* **27**, 303–351.

Wallraff, H. G. 1971. Kurzzeitige Schwankungen bei der Richtungswahl abfliegender Brieftauben. *J. Ornithol.* **112**, 396–410.

Wallraff, H. G. 1972a. Homing of pigeons after extirpation of their cochleae and lagenae. *Nature (London), New Biol.* **236**, 223–224.

Wallraff, H. G. 1972b. Nichtvisuelle Orientierung zugunruhiger Rotkehlchen (*Erithacus rubecula*). *Z. Tierpsychol.* **30**, 374–382.

Wallraff, H. G. 1972c. Fernorientierung der Vögel. *Verh. Deut. Zool. Ges.* **65**, 201–214.

Wallraff, H. G., and Graue, L. C. 1973. Orientation of pigeons after transatlantic displacement. *Behavior* **44**, 1–35.

Watson, J. B., and Lashley, K. S. 1915. An historical and experimental study of homing in birds. *Carnegie Inst. Wash., Publ.* **1**, 7–60.

Wehner, R., and Labhart, T. 1970. Perception of the geomagnetic field in the fly *Drosophila melanogaster*. *Experientia* **26**, 967–968.

Wenzel, B. M. 1971a. Olfaction in birds. *In* "Handbook of Sensory Physiology IV. Chemical Senses 1, Olfaction" (L. B. Beidler, ed.), pp. 432–448. Springer-Verlag, Berlin and New York.

Wenzel, B. M. 1971b. Olfactory sensation in the Kiwi and other birds. *Ann. N.Y. Acad. Sci.* **188**, 183–193.

Whiten, A. 1972. Operant study of sun altitude and pigeon navigation. *Nature (London)* **237**, 405–406.

Whitney, L. F. 1969. "The Basis of Breeding Racing Pigeons." Faber & Faber, London.

Williams, T. C., Williams, J. M., Teal, J. M., and Kanwisher, J. W. 1972. Tracking radar studies of bird migration. *In* "Animal Orientation and Navigation." NASA SP-262, pp. 115–128. U.S. Gov. Printing Office, Washington, D.C.

Wiltschko, W. 1968. Über den Einfluss statischer Magnetfelder auf die Zugorientierung der Rotkehlchen (*Erithacus rubecula*). *Z. Tierpsychol.* 25, 537–558.

Wiltschko, W. 1972. The influence of magnetic total intensity and inclination on directions preferred by migrating European Robins (*Erithacus rubecula*). *In* "Animal Orientation and Navigation." NASA SP-262, pp. 569–578. U.S. Gov. Printing Office, Washington, D.C.

Wiltschko, W., and Höck, H. 1972. Orientation behavior of night migrating birds (European Robins) during late afternoon and early morning hours. *Wilson Bull.* 84, 149–163.

Wiltschko, W., and Merkel, F. W. 1966. Orientierung zugunruhiger Rotkehlchen im statischen Magnetfeld. *Verh. Deut. Zool. Ges. Jena, 1965* pp. 362–367.

Wiltschko, W., and Merkel, F. W. 1971. Zugorientierung von Dorngrasmücken (*Sylvia communis*) im Erdmagnetfeld. *Vogelwarte* 26, 245–249.

Wiltschko, W., and Wiltschko, R. 1972. Magnetic compass of European Robins. *Science* 176, 62–64.

Wiltschko, W., Höck, H., and Merkel, F. W. 1971. Outdoor experiments with migrating European Robins in artificial magnetic fields. *Z. Tierpsychol.* 29, 409–415.

Windsor, D. M. 1972. Directional preferences and their relation to navigation in homing pigeons (*Columba livia*). Ph.D. Thesis, Cornell Univ., Ithaca, New York.

Wodzicki, K., and Wojtusiak, R. J. 1934. Untersuchungen über die Orientation und Geschwindigkeit des Fluges bei Vögeln. I. Experimente an Schwalben (*H. rustica* L.). *Acta Ornithol. Mus. Zool. Pol.* 1, 253–274.

Wodzicki, K., Puchalski, W., and Liche, H. 1938. Untersuchungen über die Orientation und Geschwindigkeit des Fluges bei Vögeln. III. Untersuchungen am Störchen (*Ciconia c. ciconia*). *Acta Ornithol. Mus. Zool. Pol.* 2, 239–258.

Wodzicki, K., Puchalski, W., and Liche, H. 1939. Untersuchungen über die Orientation und Geschwindigkeit des Fluges bei Vögeln. V. Weitere Versuche an Störchen. *J. Ornithol.* 87, 99–114.

Yeagley, H. L. 1947. A preliminary study of a physical basis of bird navigation. *J. Appl. Phys.* 18, 1035–1063.

Yeagley, H. L. 1951. A preliminary study of a physical basis of bird navigation. II. *J. Appl. Phys.* 22, 746–760.

The Ontogeny of Behavior in the Chick Embryo

RONALD W. OPPENHEIM

NEUROEMBRYOLOGY LABORATORY
DIVISION OF RESEARCH
NORTH CAROLINA DEPARTMENT OF MENTAL HEALTH
RALEIGH, NORTH CAROLINA

Dedicated to the memory of
Daniel S. Lehrman (1919–1972)
whose interest in this work was an inspiration to many

I. Introduction

With a few major exceptions (e.g., Zing-Yang Kuo, Leonard Carmichael) psychologists have rather consistently demonstrated an inexplicable reticence to study behavioral development during the embryonic or prenatal period. Although one is reluctant to attribute to psychologists the view that behavior really only begins after birth or hatching, the evidence, nevertheless, leads almost inexorably to just such a conclusion. In part, at least, this apparent indifference to the prenatal period can probably be traced historically to the intense, indeed, almost myopic interest that psychologists have devoted to adult learning mechanisms.

Biologists, on the other hand, have always been more keenly aware of the necessity for studying embryonic development in order to adequately understand adult characteristics, be they biochemical, morphological, or functional, including behavior. This difference between psychologists and biologists has in the past often led to an ignorance or misunderstanding by psychologists of the basic principles which have guided those biologists interested in early neural and behavioral development (e.g., neuroembryologists). Indeed, at least some of the blame for the continuing controversy over the role of genes and the environment in behavioral development can, I believe, be attributed to this misunderstanding.

In the present essay I want to review the current state of knowledge of embryonic behavioral development in bird embryos, including our present understanding of the neurobiological mechanisms underlying this development, and thereby try to clarify for psychologists why a thorough examination of neural and behavioral development prior to birth or hatching is at least as important as knowledge of any other developmental period for attaining a better and fuller understanding of later behavior. The choice of using the bird embryo as a model—not necessarily of how embryonic behavior develops in all vertebrates, but of how embryonic behavioral development can be fruitfully studied analytically—was motivated by several considerations: (a) birds are amniotes, lacking a larval stage of development and, therefore, more similar to higher vertebrates during their embryonic development than are fish or amphibians, (b) bird embryos are relatively accessible for analytical study over the entire period of embryonic development, and (c) because of this accessibility there is perhaps more known about embryonic behavioral development and the related neurobiological mechanisms in birds than in any other vertebrate.

On the final page of his book *Principles of Development,* Paul Weiss (1939), one of the foremost neuroembryologists of our time, made a strong plea for a systematic examination of the development of behavior and neural function in the embryo. Although it has taken more than 20 years for this plea to begin to be answered in earnest, the answers are coming primarily from biologically trained investigators. Yet, there are a multitude of questions concerning embryonic behavioral development that could be just as fruitfully attacked by psychologists knowledgeable in the techniques, concepts, and historical forebears of modern developmental biology and psychology. Perhaps the present essay will serve to generate an interest in some of these questions.

The chick embryo has already begun its tortuous journey from zygote to hatchling before the egg is even laid (Patten, 1958). A discussion of the tremendously complex, but extraordinarily interesting, events and mechanisms that are responsible for this developmental process, however, are beyond the scope of the present essay. Nevertheless, it is important for the reader to be aware of the fact that the chick embryo, having begun its life outside the hen with only a few undifferentiated cells, is transformed in the short period of 20 days into a highly organized precocial organism, that within a few hours after hatching, is almost entirely capable of fending for itself (Nice, 1962). Concerning the problem of the organization of behavior and the nervous system, this transformation raises questions that surely are of direct relevance to all of developmental psychology. Let us consider then the extent to which the chick embryo has allowed us to reveal a few of its well-kept secrets.

II. Chronology of Behavioral Events during Incubation

A. The Early Phase (Day 3 to Day 9)

Beginning sometime between the third and fourth days of incubation very slight unilateral contractions of the cervical neck musculature can be observed in the domestic chick embryo. These first feeble signs of overt behavior may, depending upon their strength, actually cause the head of the embryo to move slightly to one side or the other. At first the movements occur rather infrequently (i.e., about 0.5–0.8 times per minute according to Foelix and Oppenheim, 1973), but gradually they increase in number. Of course these movements cannot be seen in the unopened egg, even with strong transillumination; it is necessary to make a window in the shell and underlying membranes so that the embryo can be observed under the microscope. This is a simple technique which, if done properly, allows observation of the relatively undisturbed embryo throughout the entire incubation period (Oppenheim et al., 1973). Very soon after their inception, the movements begin to occur alternately with contractions of the musculature occurring on one side of the midline, and then on the other, with a rather short latency of about 0.5 seconds. A few hours later these side-to-side movements begin to spread from the neck region down the trunk so that by about 5 days even the tail is involved. Concomitant with this caudal spread of activity, the side-to-side movements can be seen to occur more rapidly and several such undulating or S-wave movements may occur in quick succession. Initially only the neck and trunk musculature exhibits active contractions; the fore- and hindlimbs are only *passively* or mechanically moved by the trunk contraction at this time. By 6 days, however, *active* flexions and extensions of the limbs occur; at first only at the same time as general trunk movement, but within 24 hours they may occur independently. The frequency of this general trunk and limb activity continues to increase so that

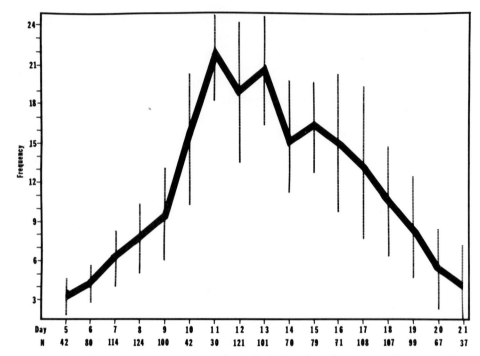

FIG. 1. Frequency (per minute) and standard deviation of all observable embryonic movements of the chick throughout the incubation period.

by 7–8 days there are about 6–7 movements per minute (Fig. 1). In many respects the undulating or S-wave movements occurring during this early phase resemble a slow-motion version of the swimming pattern of fish.

Perhaps the most striking characteristic of these early movements is their temporal pattern. There is a regular periodicity in the occurrence of the movements such that a burst of one or several movements occurs, followed, after a quiescent period, by a second burst of movement, and so on (Fig. 2).[1] This cyclic pattern exhibits a specific rhythm at each stage of develop-

[1] This rhythm is roughly defined by the average duration of *activity* period or burst, and the average duration of the *inactive* period separating two bursts of activity, and is referred to as *cyclic motility* (see Hamburger and Balaban, 1963; Hamburger *et al.*, 1965). Typically, any period lasting 9–9.5 seconds during which no movements occur has been arbitrarily defined as an inactivity period. Shorter periods of quiescence are treated as activity and are summed together with the contiguous activity period. Although this measure fairly accurately reflects the rhythmic nature of the embryo's movements during the early stages it tends to become more artificial as the embryo gets older, particularly when it is used as a measure of the total activity of an embryo during a given recording period (Oppenheim, 1970).

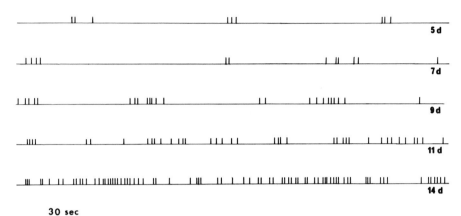

FIG. 2. Actual segments from motility recordings of chick embryos at various developmental stages. Each vertical line represents a discrete movement of some part of the embryo. Note particularly the changes in the temporal pattern and the frequency of movement with age.

ment until about 11–12 days of incubation, at which time the movements become irregular. Although this rhythmicity in the chick embryo's movement had been reported by several early investigators (Clark and Clark, 1914; Visintini and Levi-Montalcini, 1939; Rhines, 1943), Hamburger (1963) was the first one to clearly recognize the significance of this aspect of the embryo's behavior (see Section IV,A).

In summary then, the first phase of embryonic behavior in the chick is characterized by: (a) the spread of active muscular contractions from the cervical neck region to the trunk, and then to the limbs; (b) a gradual increase in the frequency of movements; (c) a preponderance of rather stereotyped S-wave or swimming-like movements involving the entire trunk musculature; and (d) a regular periodicity of the movements.

It is important to note at this point that in the present discussion of the three phases of behavioral development in the chick embryo I am limiting my attention to the *spontaneous* movements of the embryo, that is to all movements not elicited by deliberately applied experimental stimulation.

B. THE MIDDLE PHASE (DAY 9 TO DAY 16)

The choice of dividing points between the three phases of embryonic behavior in the chick is rather arbitrary (as is also the choice of three rather than five, ten, etc., phases), and this is nowhere better illustrated than in the transition from the early to the middle or second phase.

Beginning around 9 days and continuing up to 11–12 days of incubation the periodicity of the embryonic movements *gradually* becomes more arrhythmic (Fig. 2). Even earlier than this (by days 6–7) the stereotyped and coordinated S-wave movements begin to be replaced by more irregular, unintegrated jerky movements of many different parts of the embryo (Orr and Windle, 1934; Hamburger, 1963). These latter movements we have termed Type I and Type II activities (Hamburger and Oppenheim, 1967). The Type I activity appears to be the random summation or combination of movements of individual parts, such as the head, trunk, limbs, beak, etc. They are not smoothly executed as is the case of Type III activity (see below) and posthatching motor activity. Type II activity consists primarily of sudden startles, involving an almost synchronous activation of the entire neuromuscular system. They are typically of brief duration, but of rather high amplitude. It is also during this transition period that more and more parts of the embryo become active. For example, between days 6 and 12, independent limb and digit activity, beak movements, eyelid movement, etc., are all seen for the first time (Kuo, 1932a; Hamburger and Balaban, 1963; Kaspar, 1964). Therefore, although this *middle phase* of behavioral development in the chick embryo is primarily characterized by (a) the lack of clear temporal periodicity, and (b) a relative decrease of coordinated activity (the S-waves of phase 1), the final attainment of these characteristics is a gradual process that isn't completed until about the fourteenth or fifteenth day of incubation. However, during this second phase of behavioral development the frequency of movements reaches a peak of about 20–25 per minute, then it sharply declines (Fig. 1). A satisfactory description of the different types of embryonic movements that occur during this phase, as well as their relative chronological appearance, while beyond the scope of the present discussion, can be found in Kuo (1932a) and Kaspar (1964).

During phase 1 (early), as well as during much of the transition from the *early* to the *middle* phases, independent movements of separate parts of the body are infrequent; when the embryo is active it appears as if almost the entire functional neuromuscular system is involved (see Section IV,C for a possible neurophysiological explanation of this). Phase 2, however, is characterized by an increased frequency of independent movements (Kuo, 1932b,c; Oppenheim, unpublished observations). Although much of the overt behavior of the embryo during phase 2 is best characterized as jerky, unintegrated, almost convulsivelike activity, there do occur occasional, but rare instances of integrated activity of several parts of the body: alternating flexions and extensions of the legs resembling locomotory movements, simultaneous flapping of the wings, and sequential beak opening and closing and swallowing are some examples. In spite of these few instances of integrated behavior, however, it is generally true that in going from phase 1 (early) to phase 2 (middle) there is a change from a predominance of rather stereotyped, simple integrated movements—the S-waves—to movements which are much less stereotyped and seemingly less integrated (Types I and II).

C. The Final or Hatching Phase (Day 17 to Hatching)

The general trend described above continues until shortly before hatching, when it begins to be interrupted for short periods of time around the seventeenth day of incubation. At the beginning of this *final* or hatching phase, the movements can be briefly characterized in the following manner: Behavioral bouts or episodes occur which may last for several minutes; qualitatively the movements are stereotyped in form and tend to follow one another in a rather regular temporal pattern. These coordinated movements, which have been labeled Type III or rotatory tucking movements (Hamburger and Oppenheim, 1967), are involved in the initial preparation of the embryo for emergence from the shell. They serve to shift the embryo into a position so that the beak can penetrate the membranes in the air space and the embryo can begin lung ventilation and later pip or crack the shell. This new behavior pattern continues sporadically with slight variations in form, during the remaining 3–4 days of incubation, and up to the moment of actual emergence from the shell. In fact it is one variant of this Type III pattern that is responsible for hatching and emergence, when the embryo cracks the shell around the circumference, pushes the shell cap off, and wriggles out of the egg. An important feature of these Type III movements which, like their coordinated nature, distinguishes them from the more unintegrated, and convulsivelike Type I and II activity occurring during this time, is their temporal pattern. They occur in bouts or episodes during which the other more uncoordinated Type I and II movements are almost entirely absent. Within each episode the Type III movements also occur rather regularly about once every 10–20 seconds (Hamburger and Oppenheim, 1967; Kovach, 1970; Corner and Bot, 1967). Almost immediately after the termination of such an episode the more jerky, uncoordinated Type I and Type II movements reappear and continue until the next episode of the Type III activity. It is possible that the long cycles of variation in the frequency of embryonic movement in the chick (about 2–3 hours) reported by Tremor and Rogallo (1970) may reflect the alternation between these two basic types of motor activity. The frequency of movement per unit time is certainly decreased during one of these Type III bouts (Kovach, 1970).

In this last phase of behavioral development Type I and Type II movements consist increasingly of low-amplitude jerks and twitches of isolated parts (e.g., slight toe movements or head jerks). It is also during this time that the frequency of oral activity known as beak-clapping increases (Hamburger and Oppenheim, 1967). In summary, beginning about mid-way through the middle or second phase of behavioral development (about 13–14 days) Type I and Type II movements decline in frequency, and more and more of this activity consists of low amplitude movements of separate parts. On about the seventeenth day of incubation a new integrated motor pattern, Type III, first manifests itself. The alternation between these two basically different types of behavior continues until 1–2 hours prior to the chicks' emergence from the shell, at which time the Type III pattern becomes dominant. Between 17 days and the onset of this brief hatching period, the Type III movements comprise only 5–10% of the total activity observed, whereas, during the actual hatching process they occupy more than 90% of the observed activity. It is during this final 1 to 2 hour hatching period (which we have called *climax*) that the embryo breaks the shell cap off and emerges.

This then concludes my description of the embryogenesis of the spontaneous or unsolicited behavior of the chick embryo. From observations of other species of bird embryos, including altricial, as well as precocial forms (Tuge, 1937; Kaspar, 1964; Oppenheim, 1970, 1972a, 1973; Harth, personal communication), it appears as if the basic facts described here for the chick also apply. Furthermore, distantly related forms, such as fish (Tracy, 1926), amphibians (Hughes, 1966), and reptiles (Decker, 1967; Hughes *et al.*, 1967) also exhibit some of the same behavioral characteristics as the chick during their embryonic development. The extent to which the behavior of precocial and altricial mammalian embryos and fetuses adhere to the general features observed in the chick, such as periodicity and spontaneity, is difficult to determine, since most of the mammalian studies have been done under rather unfavorable environmental conditions, and because the investigators were seldom concerned with spontaneous, nonstimulated behavior (Hamburger, 1963). In spite of these difficulties, however, it is clear from the literature that all mammalian embryos that have been observed do exhibit some spontaneous movement (Hamburger, 1963, 1970, 1971). Indeed, a recent reinvestigation of embryonic movements in the rat embryo has revealed several important similarities between the rat and chick (Narayanan *et al.*, 1971). For example, both forms display: (a) a periodicity in their spontaneous motility; (b) a general lack of integration in the movements of parts; and (c) a gradual buildup of spontaneous activity, reaching a peak on day 12–13 in the chick and on day 18 in the rat, with a subsequent sharp decline in activity prior to birth or hatching. The one major difference between the chick and rat is that the rat does not exhibit a prereflexogenic period of motility; both spontaneous and reflexively evoked movements arise at the same time. This also appears to be true of other mammals (Hamburger, 1963), including higher primates (Bodian, 1968). These sparse findings underline the urgent need that exists in the field of behavioral embryology for a careful reexamination of the behavioral repertoire and development of undisturbed mammalian embryos under more favorable conditions of maintenance than have been used in the past.

III. The Ontogenetic Organization of Motor Behavior

One of the most difficult questions facing students of early behavioral development is how the investigation of embryonic and fetal movements, and their neurobiological correlates, can help us better understand the complex action patterns performed by the newborn or newly hatched organism. Or stated somewhat differently, what are the governing rules or mechanisms whereby the spontaneous and reflexogenic movements of the embryo later become organized into functionally adaptive sequences of integrated behavior. There are at least four possible answers to this question.

First, there may be no relationship at all. The behavior and neurobiological mechanisms operating at one stage of development may be transitory or provisional, and of adaptive value only for that particular stage. I am aware, of course, that such a possibility would violate the well-accepted tenet of developmental biology which states that every step in structural and functional neural differentiation emerges from the preceding steps in a gradual and progressive fashion. Yet there appear to be some notable exceptions to this rule. For example, the human neonate exhibits several reflexes, as well as more complex responses, which are subsequently lost and whose relationship to later behavior is still in question (e.g., see McGraw, 1939; Ausubel and Sullivan, 1970). In all vertebrates there is, in many parts of the nervous system, an overproduction of neurons which subsequently degenerate (Hamburger and Levi-Montalcini, 1950). In some cases these cells have already begun axonal outgrowth and, indeed, may even have established morphological connections with other nerve cells or with their normal peripheral field (Hughes, 1968; Prestige, 1970; Prestige and Wilson, 1972). Although it is still an open question whether such neurons, which will subsequently die, can form functional connections, there is some evidence in support of this suggestion (e.g. Bagust, Lewis, and Westerman, 1973). Thus, it seems rather likely that in many different parts of the nervous system neurons are produced, which may differentiate and initiate function, only to subsequently die through some as yet unknown mechanism. If such neurons play any role in the mediation of embryonic motility as described here for the chick, one would have an extreme case of discontinuity, at least on the structural level. Of course, other such extreme examples are not uncommon; in both invertebrates and most amphibians metamorphosis may be accompanied by the death of neurons which formerly played a functional role during embryonic and larval life (e.g. Mauthner cells, Rohon-Beard cells). These few examples suggest that on both the behavioral and neural levels it may not be uncommon for a sharp discontinuity to exist between different ontogenetic stages; so that even by knowing in detail the behavioral repertoire or neural structure at one stage we still would not be able to predict or better understand subsequent stages. Admittedly such discontinuities may not characterize a majority of neurogenetic events, but it may well be applicable to some special cases and therefore should be considered as a viable alternative to the original question posed above.

A second alternative is that while there may be a functional relationship between the movements of different parts of the embryo at two or more progressive stages of development the level of our analysis, or the imprecision of our techniques, could be insufficient to detect such a relationship. For example, in the chick embryo alternating leg movements are a rare occurrence until after hatching when the chick actually begins to locomote. This could represent a discontinuity of the kind I discussed above, or alternatively it is possible that by recording electromyographically from various muscles in the two legs at different stages one could detect changes that were of a more gradual, progressive nature leading up to true postnatal locomotion.

Still another alternative which has been recently discussed by Hamburger (1973) is that in some instances there may only be discontinuity on the behavioral level. For example, the Type III pattern performed by the chick during prehatching and hatching stages apparently does not arise as a *gradual* organization of the earlier Type I and Type II movements (Hamburger and Oppenheim, 1967), even though the neuromuscular mechanisms may be the same. Now this finding, if correct, presupposes one of two things: either the Type I and Type II movements are organized into the Type III pattern by influences from parts of the central nervous system (CNS) that only become functional rather suddenly on day 17 (see Section IV,B); or the neural mechanisms mediating the Type III pattern are entirely different from those involved in the Type I and Type II activity. In either case, it is necessary to assume—if, indeed, there is continuity on the structural or physiological level—that those mechanisms underlying Type III behavior have been undergoing gradual and progressive changes throughout development, but that the embryos' *overt* activity does not faithfully reflect these neurogenetic events going on "under the surface," perhaps due to powerful active inhibitory mechanisms (Crain, 1973).[2]

---

[2] The precocial sheep fetus is apparently almost completely quiescent during the final 40 days of gestation, due to inhibitory influences from the brain (Barron, 1941). Yet, within minutes after birth it performs many complex action patterns. If one adheres to the notion of a gradual continuity in *overt* behavioral development then it is necessary to attempt to understand postnatal behavior by reference to overt antecedents that last occurred 40 days previously.

Finally, there may, in fact, be a functional relationship between the overt movements at different stages of development. If this is the case, however, it is necessary to make a clear distinction between the two possible ways by which such a functional relationship may work.

In the first place early movements could be partial approximations of later more complex behavior. For example, swimming in fish and amphibian larvae seems, on the descriptive level, to be gradually built up from the simpler unilateral and bilateral trunk contractions of the embryo, as described by Coghill (1929). This would merely represent a continuity or functional relationship in the *form* of behavior. On the other hand, to use the same example, the overt performance of the early partial movements could not only be approximations of the swimming pattern but they might also be a necessary condition for the normal manifestation of swimming. In this case, elimination or suppression of these early simpler movements—and the sensory experiences accompanying them—should lead to a deficit in the later behavior, whereas in the first type of functional relationship it would not. In other words, the crucial characteristic distinguishing these two types of functional relationships is the necessity of the overt occurrence of the early movements for the normal manifestation of the later pattern. Although there are very few analytic studies that bear on this question, the consensus has been that suppression of early embryonic activity by pharmacological immobilization does not impair the performance of the later behavior (Harrison, 1904; Matthews and Detwiler, 1926; Carmichael, 1926, 1927, 1928; Fromme, 1941). Because many fish and amphibian embryos hatch at an incomplete larval stage of development, but must fend for themselves at this time, it is perhaps not surprising that they exhibit such a straightforward behavioral development which is highly buffered against environmental (i.e., sensory experience) contingencies. So far similar experiments have not been carried out *in vivo* with higher forms.

Even on the descriptive level, however, embryonic behavioral development in amniote forms, especially precocial species, is not nearly as clear-cut as in fish or amphibians. For example, the newly hatched chick exhibits a multitude of complex coordinated action patterns, such as walking, pecking, perching, preening, etc. (Bangert, 1960; Kuhlmann, 1909; Nice, 1962; Schooland, 1942; Simon, 1954; Spaulding, 1872), whose occurrence would have been practically impossible to predict even with a detailed description of embryonic behavior during the entire preceding incubation period.

Indeed except for the early S-waves—which is a transitory behavior whose pattern is probably dictated by specific neuroanatomical constraints in the spinal cord—and the rare appearance of alternating leg movements, bilateral wing-flapping, and Type III movements, there are few instances in which the chick embryo performs coordinated movements involving different parts;

and in these cases it has not yet been possible to trace a gradual or continuous buildup of the *form* of these patterns from earlier movements. Even if such a relationship should be found, however, we would still be faced with the question of *causal* relationships between the different stages.

No discussion of the embryonic antecedents of behavior would be complete without reference to the pioneering efforts of the psychologist Zing-Yang Kuo. In a long series of papers published during the 1930's Kuo attempted to remove some of the mystery surrounding the concept of innate or instinct by studying the embryonic development of behavior in the chick.

From the start, Kuo (1922) was interested in determining how the so-called early "random" movements "are integrated and reintegrated into various reaction systems . . ." (p. 353) during ontogeny. It was his contention that this integration resulted almost exclusively from a continuous process of interaction between an active organism and its environment (but it is important to realize that, according to Kuo, the embryos' activity was caused or stimulated by sensory input and was not occurring spontaneously). In other words, the *overt* activity of the embryo was viewed as a necessary and important component in the causation of later organized behavior (see Oppenheim, 1972b). Obviously the paralysis experiments discussed above were in direct opposition to this view. It simply was not conceivable to Kuo that complex patterns of behavior could develop without the prior occurrence of sensory initiated and guided movements (Kuo, 1932d,e).

Although recent deafferentation experiments with the chick embryo (see Section V,B) have shown that certain aspects of behavior can develop normally without the embryo having ever experienced sensory input, the normality of Type III hatching behavior or posthatching behavior was not systematically examined in those experiments. Therefore, Kuo's contention that sensory input during ontogeny is the primary determinative factor in the organization of *later integrated behavior* remains to be tested in the chick.

I have argued elsewhere—in partial agreement with Kuo—that the individual units of activity that comprise the Type III hatching pattern in the chick do not arise *de novo* the first time that pattern is performed on day 17, but that most, if not all, of these components have been occurring sporadically prior to day 17. The only novel feature that occurs on day 17 is that the units seem to become organized into a unique functional sequence (Oppenheim, 1970). Such an explanation, of course, only deals with the problem on a descriptive level and still leaves unanswered the question of the causal mechanisms (sensory input, brain influences, hormones, etc.) that may be responsible for the new behavioral integration.

If there is any single conclusion that arises from the foregoing discussion it is how meager our knowledge is of the mechanisms and rules that govern the ontogenetic organization of muscle activity into functionally adaptive motor action patterns. Although it is still too early to decide the importance of either sensory input or motor activity occurring during the embryonic period in this organization, when the final verdict is read, I am convinced that such influences will be found to be an integral part of this process, even during the earliest stages of embryogenesis.

IV. Neurobiological Events Underlying the Embryogenesis of Behavior

A. Neurogenic versus Myogenic Basis of Embryonic Behavior

One of the first and most basic questions that must be answered in the study of embryonic behavior is whether the movements (both spontaneous and elicited), particularly during early stages, are actually mediated by the nervous system (neurogenic), or whether they are due solely to nonneural contractions of muscle tissue (myogenic). Certain fish, for example, exhibit spontaneous and rhythmic patterns of movement that have been shown to continue unaffected after removal of the entire spinal cord (Wintrebert, 1930), and are therefore truly myogenic. This question is particularly difficult to answer on neuroanatomical grounds alone in the case of the chick embryo because, although motor nerves are already contiguous with muscle tissue when the first movements occur on day 3.5 (Visintini and Levi-Montalcini, 1939), mature neuromuscular junctions are not formed until the tenth to twelfth day of incubation (Drachman, 1963; Hirano, 1967; Mumenthaler and Engel, 1961). Furthermore, it has long been known that isolated embryonic muscle when placed in tissue culture contracts rhythmically and spontaneously (Lewis, 1915).

Several approaches have been used in an attempt to settle this question in the chick embryo. The first, and perhaps least reliable, is the use of a neuromuscular blocking agent, such as curare. In the adult organism curare serves to block the transmission of CNS bioelectric activity to striated muscle, therefore the continuation of movement after the appropriate application of curare is generally taken as evidence of myogenic activity. The question arises however whether the embryonic neuromuscular junction responds to curare in the same way as that of the adult (Barron, 1941). In certain fish, at least, embryonic and adult muscle do appear to respond in a similar fashion to curare (Brinley, 1951). Although similar information is not available for birds it is known that from their inception all movements in the chick embryo are blocked by curare (Levi-Montalcini and Visintini, 1938a; Kuo, 1939a; Hamburger, 1963). Tuge (1937) on the other hand, has reported that curare does not eliminate all movement in early pigeon embryos, and thus he has concluded that, similar to fish, the pigeon embryo exhibits an initial myogenic phase of activity. We have recently been unable to replicate this finding and therefore it seems likely that Tuge may not have distinguished between active embryonic movement and passive movement caused by the swinging amnion. Indeed, according to Tuge's description, the movements remaining after curare injection resemble the *passive* embryonic movement we have observed.

A different and somewhat more direct test of this question has been reported by Alconero (1965). She excised pieces of muscle tissue (somites) from chick embryos and transplanted them onto the highly vascularized chorioallantoic membrane of host embryos, where they were well nourished and continued development. In some of the cases adjacent neural tissue (spinal cord) was included in the transplant, and in the others only muscle tissue was used. She found that only those transplants that contained neural tissue which had actually innervated the muscle exhibited spontaneous contractions; the neurally isolated muscle tissue was never seen to contract. Although this study is subject to the criticism that the chorioallantoic environment may not have been adequate to allow myogenic activity in the isolated muscle transplants, these findings, together with the curare data, nevertheless lend support to the idea that the chick embryo lacks a true myogenic phase of behavioral development.

Still another approach to this question is that of Levi-Montalcini and Visintini (1938b). They reported that electrical stimulation of the 4- to 5-day-old chick embryo's brain produced overt movements that were indistinguishable in form from spontaneous movement, again apparently supporting a neurogenic explanation. Inadequate control of the electrical stimulus, however, renders these findings inconclusive, for the stimulus may have spread to the musculature thereby causing direct muscle contractions. More recently, Ripley and Provine (1972) have demonstrated unequivocally that embryonic movements, even as early as 4 days of incubation, are almost always accompanied by, and closely correlated with, bioelectric activity in both the spinal cord and the peripheral motor nerves. When this finding is considered along with the other data bearing on this issue it now seems safe to conclude that from its inception, the spontaneous behavior of the chick embryo is mediated by the nervous system and therefore, unlike that of certain fish, does not pass through an initial myogenic phase.

B. Brain Influences in Embryonic Behavior

In this section I would like to deal with the question of when and how the brain exerts its influence on the motor behavior of the embryo. Neuroanatomically, it is still an open question of *when* fiber tracts from the brain first grow into the spinal cord. Visintini and Levi-Montalcini (1939) claim to have demonstrated the presence of the descending medial longitudinal fasciculus (MLF) tract (which originates in the mesencephalon and diencephalon) in the most caudal part of the spinal cord by 4 days of incubation. They contend that this tract is already functional at this time and together with the precocial appearance of the ventral commissure system serves as a longitudinal integrating system producing the waves of bilateral contraction that pass up and down the trunk of the embryo (S-waves). Spinal cord transection or destruction of specific brain parts, they report, results in both quantitative and qualitative changes in the embryonic movements *below* the ablation. Rhines (1943), on the other hand, was unable to confirm either these neuroanatomical or these functional findings. According to her report, the MLF tract doesn't even reach the rostral-most spinal cord until 8–9 days, and early interruption of this tract by brain ablation produces no deficits in the embryo's movements, at least up to 6 days. Since the behavioral data in both of these reports were based on only a few embryos, the question of a possible early brain influence on the embryo's movements required further examination.

We are currently investigating this problem in our laboratory. To eliminate influences from the brain we removed a segment of neural tissue in the cervical spinal cord at 2 days of incubation thereby creating a spinal "gap" across which the growing fiber tracts from the brain could not pass.[3] These embryos were allowed to develop to either $4\frac{1}{2}$ to 5, or $6\frac{1}{2}$ to 7 days of incubation at which time all the movements occurring below the "gap" were recorded for 10 minutes. The gaps were checked and confirmed histologically. In brief, we found no statistically reliable effects of the gap on the absolute frequency or the cyclic aspects of movement in either experimental group. Furthermore, no qualitative differences in the form of movement were detected at either age. Therefore, it now seems certain that if fiber tracts from the brain are normally present in the spinal cord of 4- to 7-day-old embryos they do not serve any function that can be detected in the embryo's overt behavior. The next obvious question of course is whether the brain has an influence on the embryo's behavior *after* 7 days.

Indeed, there are four separate studies that have reported such a brain influence beginning at about 8 days and continuing at least up to 17 days of incubation (Hamburger and Balaban, 1963; Hamburger *et al.*, 1965, 1966; Decker and Hamburger, 1967). In these investigations spinal cord gaps or specific brain ablations reportedly caused a *reduction* of 10–20% in the amount of movement below the operation, as well as a modification

[3] For a description of the embryonic microsurgical techniques see Hamburger (1960), Wenger (1968), and Narayanan (1970).

of the periodicity of movement. The interpretation of these results has typically been that normally the brain (or anterior spinal cord above the transection) provides excitatory input to the remaining spinal cord which serves either to prolong an ongoing activity phase or to shorten an inactivity phase by initiating additional movement. Unfortunately, because these early studies did not consider the *actual number* of embryonic movements, but instead relied upon the arbitrary and somewhat artificial measure of cyclic motility, these data do not provide conclusive support for this interpretation (see footnote 1). Therefore, it is still an unresolved question whether the brain input actually stimulates additional movements at these later stages of incubation, or whether it merely serves to modulate the temporal pattern (periodicity) of the movements.

The first indication that the actual *frequency* of movement is *not* modified by removal of potential input from the brain appeared in a study of the effect of spinal gaps upon hatching behavior (Oppenheim and Narayanan, 1968). No significant differences were found in the frequency of leg movements between control and thoracic-gap embryos at 19–20 days of incubation. Subsequently, while examining the effects of specific brain ablations upon hatching behavior, it was also found that neither midbrain nor forebrain removal modified the frequency of general embryonic movement between 17 and 20 days of incubation (Oppenheim, 1972c). Although these more recent findings all deal with rather late stages of development they do suggest that the previous interpretation of the role of the brain may have been incorrect.

Therefore, we have reexamined this question by concentrating on 10- to 11-day embryos, a stage in which marked behavioral effects after spinal gaps were shown in one of the previous studies (Hamburger *et al.*, 1965). Once again several segments of nerve tissue were removed in the cervical spinal region of 2-day chick embryos, thus prohibiting brain input to the trunk. At 10–11 days of incubation the operated ($n = 22$), and the sham control ($n = 15$) embryos were observed and their wing and leg movements recorded on an event recorder. We evaluated both the frequency and the cyclic aspects of the movements. All gaps were verified histologically. The results are summarized in Table I. The data for the temporal or periodic aspects of the movement are practically identical to those reported in a previous study (Hamburger *et al.*, 1965), which would again lead one to conclude that *normally* at this age the brain acts to *increase* the amount of movement in the trunk and appendages. Yet, the more direct test of this idea, namely, measuring the actual frequency of movement, does not support such an interpretation: no reliable difference was found in the *frequency* of movement. Similar negative results have also been found with 15-day spinal gap embryos.

On the basis of all of the foregoing I would like to suggest the following

TABLE I

SUMMARY OF RESULTS FROM CONTROL AND SPINAL GAP EMBRYOS AT
10–11 DAYS OF INCUBATION

Embryos	Number	Frequency of movement (per minute)	Total activity (%)	Activity phase (seconds)	Inactivity phase (seconds)
Control	15				
Mean		19.3	60.5[a]	32.6[b]	21.6[c]
SD		4.3	8.0	8.9	4.6
Spinal gap	22				
Mean		17.6	48.7	26.7	31.3
SD		3.1	9.7	11.1	13.1

[a] $p < 0.001$.
[b] $p = 0.05$.
[c] $p < 0.01$.

modification of the commonly held view that the brain is a source of excitation for the movements of the chick embryo. It now appears that prior to the eighth or ninth day of incubation the brain plays no role in the *overt* embryonic movements of the chick. Beginning about this time, however, and continuing up to at least 17 days of incubation, removal of the input from the brain causes a *temporal* redistribution of the movements, as evidenced by changes in the parameters of cyclic or periodic movement (i.e., the average duration of an activity or inactivity period), but it does not produce an increase or decrease in the *actual frequency or amount of movement*. And, finally, with one exception (Oppenheim, 1972c), no *qualitative* changes in movement patterns have been detected at any time after removal of brain input. The exception is that on about the seventeenth or eighteenth day of incubation absence of the midbrain produces a striking modification in the integrated Type III behavior of the chick embryo; and forebrain removal results in the failure of the embryo to initiate the final hatching and emergence phase of behavior (Oppenheim, 1972c; Corner et al., 1973). Therefore, the appearance and maintenance of these complex, integrated Type III patterns is apparently dependent upon influences from specific regions of the brain. In spite of these important exceptions, it is nevertheless true that throughout development, the overt embryonic behavior of the chick is *primarily mediated by spinal mechanisms*. Although slightly modified by brain removal, the regular rhythmic nature of the embryo's activity, as well as the normal *amount* of activity can still develop and be maintained in the complete absence of all input from the brain. This finding was rather unexpected since in sheep fetuses embryonic activity, both spontaneous and reflexive, is markedly reduced during the last 50% of the gestation period, and this inactivity has been shown experimentally to result from powerful inhibitory mechanisms located in the brain (Barron, 1941).

C. ELECTROPHYSIOLOGICAL AND NEUROCHEMICAL CORRELATES

Except for a few EEG (electroencephalograph) and evoked potential studies done during rather late stages of development (see Corner et al., 1967), the development of bioelectric activity in the chick embryo brain and spinal cord, has until quite recently received virtually no attention. Within the last 3 years, however, a series of papers has appeared in which electrophysiological recordings have been systematically made at various specific regions in the spinal cord of the chick embryo throughout the entire incubation period.

From its beginning this work had as its primary goal the search for bioelectric neural correlates of overt embryonic behavior (Provine et al., 1970; Provine, 1972; also see Bursian, 1969). Briefly, the technique consists of recording with microelectrodes the spontaneously occurring *polyneuronal burst discharges* that result from the concerted firing of many neural units. Recordings are made at many different sites and levels of the spinal cord, with primary concentration on the ventral two-thirds of the cord, containing the inter- and motoneurons. Recordings are now available for practically all stages of behavioral development from 4½ to 20 days of incubation. Furthermore, although most of the data have been obtained from curarized embryos, recordings have also been made of noncurarized embryos so that a *direct* correlation can be made between overt behavior and the spontaneous bioelectric activity.

The characteristic changes that occur in individual bioelectric burst discharges (and unit activity) during embryogenesis are depicted in Fig. 3. The discharges begin as simple accelerations in unit firing at 5 days and progress to a more complex configuration by 7–8 days. In later stages the bursts become longer, more frequent, and more regular in their temporal pattern. This long-term regularity in the periodicity or temporal pattern of the bursts can be clearly seen in Fig. 4. By determining the overall percentage of a given recording period during which bioelectric burst discharges occur and comparing this value with the actual frequency of overt embryonic movements throughout incubation one begins to see that the burst discharges are an accurate reflection of overt behavior (Fig. 5). However, an even more dramatic demonstration of this relationship can be seen in Fig. 6 (Ripley and Provine, 1972). Here the actual movements of the noncurarized, freely moving embryo were recorded *simultaneously* with the recording of bioelectric burst discharges. It is interesting that although the movements

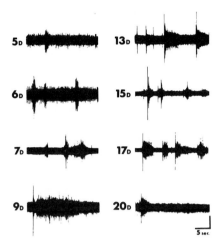

FIG. 3. Changes over time in the configuration of electrophysiological burst activity from chick spinal cord. D = day of incubation. (From Provine, 1972.)

of many parts of the embryo were observed in this study, the bioelectric activity was only recorded from a single region in the lumbosacral cord. This suggests that the bioelectric burst activity occurring at any single level of the spinal axis will accurately reflect the overt behavior occurring at several other spinal levels. It further suggests that the bioelectric bursts spread indiscriminately up and down the spinal cord, as had been previously postulated by Hamburger (1963). Indeed, Provine (1971) has demonstrated that practically simultaneous burst discharges occur at many different levels and sites within the ventral two-thirds of the spinal cord (Fig. 7). This synchronization of polyneuronal burst discharges over a wide spatial distribution is fairly accurately reflected in the embryo's overt behavior; during early stages (early phase, see Section II,A), waves of bilaterally and longitudinally integrated motility pass up and down the embryo. These stereotyped waves later disappear, but they are replaced by a more generalized irregular motil-

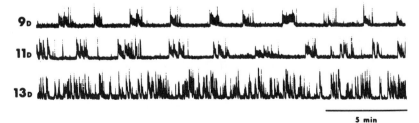

5 min

FIG. 4. Periodicity of electrophysiological burst discharges. Regular periods of activity and inactivity are present at 9 and 11 days. By 13 days the periodicity is lost. (From Provine, 1972.)

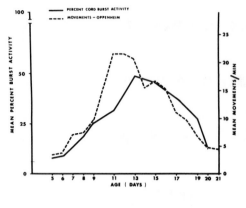

FIG. 5. Solid line represents the percentage of time during which polyneuronal burst discharges were present at different stages of development. Broken line is the mean frequency of embryonic movement throughout incubation. (From Provine, 1972.)

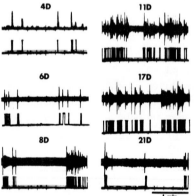

FIG. 6. Comparison of cord burst discharges (upper trace) with visually observed body movements (lower trace). Four-day cord activity was integrated to emphasize the low amplitude activity. Cord discharges were recorded from lumbosacral region except at 4 days when brachial cord was monitored. (From Ripley and Provine, 1972.)

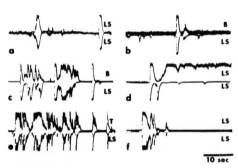

FIG. 7. Records of integrated polyneuronal burst discharges simultaneously recorded from pairs of spinal cord sites. Activity from one region is inverted and placed base-to-base with activity from the other region so that symmetry of the resulting composite trace indicates the similarity of the activity from the two sites. Correspondence (symmetry) is shown between burst discharges appearing in various pairs of spinal cord loci recorded from embryos of the following ages: (a) 6-day, bilateral, lumbosacral/lumbosacral (LS/LS); (b) 6-day, ipsilateral, brachial/lumbosacral (B/LS); (c) 9-day, ipsilateral, brachial/lumbosacral (B/LS); (d) 9-day, bilateral, lumbosacral/lumbosacral (LS/LS); (e) 17-day, ipsilateral, thoracic/lumbosacral (T/LS); (f) 20-day, bilateral, lumbosacral/lumbosacral (LS/LS). The 9-day bilateral, LS/LS case (d) shows alternating region specific activity in the two cord halves after a common initial discharge. (From Provine, 1971.)

ity in which many different parts of the embryo are active *simultaneously* (*middle phase*). These electrophysiological findings suggesting widespread integration within the cord are supported by recent ultrastructural studies of synaptogenesis. Oppenheim and Foelix (1972) and Foelix and Oppenheim (1973) have reported the presence of nascent synapses in the ventral half of the cervical spinal cord of 3- to 4-day chick embryos and in the lumbosacral region of 5-day embryos. In later stages individual synapses become more mature, and the number and types of synapses increase. Thus even at these early stages there is already a neuroanatomical basis for the kind of widespread interneuronal communication suggested by the bioelectric and behavioral data.

In certain respects at least, these studies of the ontogeny of bioelectric activity provide a more compelling example of a neural correlate of behavior than the classic studies of Coghill (1929) on the salamander. Coghill, it will be recalled, relied exclusively upon neuroanatomical indices (i.e., formation of synapses), whereas this more recent work by Provine utilizes physiology which is one step closer to the actual behavior. In other respects, however, Coghill's work was more informative in that it dealt with *specific* neural and behavioral events, whereas the chick work only provides a *general* neural correlate of *general* behavior. Provine has not yet dealt with the question of whether *specific* behavioral events (e.g., a unilateral versus bilateral leg movement or a wing versus leg movement) are reflected by *different* configurations of bioelectric burst activity. Finally, the knowledge that select aspects of bioelectric activity are directly correlated with overt embryonic behavior provides an opportunity to carry out certain kinds of experiments with the embryo that have heretofore been unfeasible (Provine, 1973; also see Section V,B).

Attempts to determine a specific and direct relationship between behavioral or electrophysiological development and changing levels of various neurochemical substances in the chick embryo have so far been unsuccessful (Kuo, 1939b; Rogers *et al.*, 1960; Burt, 1968; Vernadakis, 1969; Kellogg *et al.*, 1971; Giacobini *et al.*, 1970). The possible reasons for this failure are numerous but certainly the crudeness of both the behavioral measures and neurochemical techniques used in these studies is largely to blame. Generally the neurochemical data are obtained from such gross areas as the *entire* spinal cord and then compared to the general level of activity of the *entire* embryo. Furthermore, the functional and neurochemical data are seldom gathered by the same investigator. If one is serious about obtaining meaningful correlations between neurochemistry and function then it behooves one to: (a) restrict one's neurochemical analyses to precisely limited parts of the nervous system, e.g., lumbosacral spinal cord; (b) to similarly restrict one's behavioral or other functional interests to select parts of the embryo that are known to be controlled by the chosen neural structure (e.g. the legs) ; and (c) to obtain both the functional and neurochemical data on the same embryo and ideally in the same laboratory. The general failure to consider inhibitory as well as excitatory neurochemical mechanisms when attempting such a correlation could also result in a rather incomplete and distorted picture of the actual events.

Until the above points are taken into consideration neurochemical ontogenetic studies will merely provide us with more and more parametric data on changing levels of substances found in the nervous system. Although this information may be helpful in our understanding of the neurochemical events occurring *pari passu* with structural and functional neurogenesis it will be of little value in pinpointing those relevant neurochemical mechanisms which underlie functional ontogeny.

In Section III above, I discussed the possibility that one of the reasons the chick embryo does not exhibit an extensive repertoire of organized, coordinated activity may be the presence of powerful active inhibitory mechanisms. If this suggestion has any validity at all, then it is of course mandatory that some sign of functional inhibitory mechanisms be demonstrable in the embryo. The evidence presented in Fig. 1, that the frequency of embryonic movement decreases rather abruptly during the final third of the incubation period, although circumstantial, may, indeed, reflect the development of inhibitory mechanisms. A more direct test of this would be to determine when the chick embryo becomes behaviorally sensitive to those pharmacological agents that are thought to be direct antagonists of such suspected inhibitory neurotransmitters as GABA and glycine. These include strychnine, picrotoxin, and bicuculline (Curtis *et al.*, 1969; Barker and Nicoll, 1972). Recent evidence from work in our laboratory (in preparation) has suggested that active inhibitory mechanisms may indeed, be functional in the chick embryo. Beginning sometime between days 7 and 9 of incubation, strychnine, when injected into the egg (2–10 μg per egg) produces a brief, but statistically reliable increase in the frequency of ongoing embryonic movements. Until about 16 days of incubation this is the only noticeable effect of strychnine. At this time, however, the initial rise in ongoing activity after strychnine injection is followed by the appearance of typical adultlike strychnine seizures or convulsions, indicating a further maturation of inhibitory mechanisms. Both of these behavioral effects are mimicked in the electrophysiological recordings of burst activity taken from the chick spinal cord after strychnine injection. Picrotoxin has rather similar effects in the chick embryo, although the exact age of onset of sensitivity has not yet been determined. These results, when considered together with other behavioral and possible morphological indices of inhibition in the chick embryo spinal cord (Visintini and Levi-Montalcini, 1939; Foelix and Oppenheim, 1973) provide additional evidence for the existence of synaptically mediated active inhibition at least by the midpoint of incubation (i.e., 10 days). Although intracellular recordings are necessary to unequivocally demonstrate the presence of active inhibition, this has not yet been technically feasible with the chick embryo.

To return to my original question concerning the adaptive function of inhibition in the embryo, it should be said that the mere existence of inhibition is not sufficient proof that such a mechanism actually serves as a constraint on the overt appearance of *coordinated* behavior by the embryo; such inhibition could serve other functions, such as prohibiting the embryo from prematurely breaking the shell or internal egg membranes (Crain, 1973), or it might simply be an epiphenomenon of neural differentiation that is unrelated to any specific adaptive function *in ovo*. Indeed, so far we have not observed any increase of coordinated movements after treatment with strychnine.

V. The Role of Sensory Stimulation in the Embryogenesis of Behavior

A. Motor Primacy and Endogenously Produced Motor Behavior

Because several competent reviews have appeared recently dealing with the prenatal development of sensory systems (particularly audition and vision) and their role in the ontogeny of behavior (Gottlieb, 1968, 1970, 1971; Vince, 1973; Impekoven and Gold, 1973), I will limit the present discussion primarily to the cutaneous tactile, proprioceptive, and vestibular sensory systems and their possible influences on the ontogeny of motor behavior in the chick embryo.

The discovery of motor primacy in the chick embryo by Wilhelm Preyer in 1885 is by now a well-accepted fact (Hamburger, 1963). What this means is that the spinal neurons and musculature that comprise the motor or efferent system (including interneurons) differentiate and begin functioning *prior* to the spinal sensory or afferent system. Thus, although we have seen that the chick embryo is spontaneously active from the third day of incubation (Section II,A), no amount of external stimulation will elicit a reflexive response before the sixth or seventh day. It has frequently been implied (Hamburger, 1963; Corner and Bot, 1967; Gottlieb, 1968) that the sole limiting neuroanatomical factor which is responsible for this lack of sensitivity is the absence of the synaptic closure of the reflex arc in the CNS. Yet, Visintini and Levi-Montalcini (1939, p. 27) have reported that it is the *simultaneous* arrival of peripheral cutaneous fibers in the skin and the central closure of the reflex arc in the spinal cord, both occurring on day 6, that are responsible. This observation, which was limited by the use of the light microscope, certainly deserves reexamination with the higher resolution afforded by the electron microscope. Regardless of the neuroanatomical reason, however, the important point here is that for the initial 3–4 days of behavioral development the chick embryo is not responsive to external stimulation. Therefore, the movements that occur during this prereflexogenic period are truly spontaneous or endogenous in that afferent sensory stimuli play *no role* in their initiation or maintenance.

At about 7 days of incubation, however, the chick embryo begins to respond to the experimental application of both light-tactile and more vigorous stimulation (pokes, flips, etc.). Initially these responses are most easily elicited from the perioral region and consist primarily of total body responses, although there are rare exceptions to both of these statements (Orr and Windle, 1934; Hamburger and Balaban, 1963; Oppenheim, 1972d). Within the next 1–2 days tactile sensitivity spreads over the entire surface of the embryo. According to both Visintini and Levi-Montalcini (1939) and Tello (1922), the onset of *proprioceptive* sensitivity probably begins on about day 10 or 11 and coincides with the simultaneous appearance of muscle spindles peripherally, and the closure of a monosynaptic reflex arc, centrally. With the appearance of sensitivity to peripheral stimulation the obvious question arises as to what role this sensitivity might play in the embryo's overt behavior throughout the remainder of the incubation period. Are the early spontaneous movements prior to day 7 completely superseded by, or merged with, the reflexogenic movements? Or, do spontaneous movements continue to be a prominent and important part of the embryo's behavioral repertoire even after the onset of reflexes? Before we examine the experimental evidence bearing on these question, however, we might first ask what are the possible sources of stimulation available to the embryo?

Throughout much of its embryonic development the chick embryo is freely suspended in fluid within the amniotic sac. This sac, which consists of smooth muscle fibers, begins to contract spontaneously on about the fifth day of incubation. Within a day or two the amniotic movements have become so vigorous that the embryo within is strongly rocked and tossed about, thus setting up a possible source of stimulation. The yolk sac, which surrounds and contacts the embryo to varying degrees at different stages (Kuo, 1932b,c), is another conceivable source of stimulation; also the embryo's own movement, either by way of proprioceptive feedback or by one part touching another (self-stimulation), is another obvious candidate; and the turning of the eggs, either by the parent in nature or artificially in the incubator, is still another source of tactile, proprioceptive, and vestibular stimulation. The moderate temperature changes occurring during the *normal* course of incubation, either in nature or in the incubator, do not appear to constitute a sufficient stimulus for embryonic activity (R. W. Oppenheim and H. Levin, in preparation). Light and sound stimuli are also probably not an important source of natural stimulation for embryonic movement, at least until rather late in development (Gottlieb, 1968). However, the recent preliminary reports that the chick embryo might be sensitive to both light and sound *prior* to the functional development of either the eye or ear (Bursian, 1965; Sviderskaya, 1967) ; and the well-established fact that amphibian embryos during certain circumscribed stages of development possess a "sensory" conducting system in the skin that is not dependent upon the presence of conventional peripheral receptors, but which nevertheless can mediate movements (Roberts and Stirling, 1971; Roberts, 1971), underscores the suggestion that one should keep an open mind concerning the kinds of sensory inputs which may influence the behavior of embryos (Gottlieb, 1971).

B. Sensory Deprivation and Augmentation Studies

Although Kuo (1967) is perhaps the most well-known adherent of the view that the vast majority of behavior during embryonic development is a result of environmental stimulation, other equally prominent investigators have also expressed this bias (e.g., Windle, 1950; Barcroft and Barron, 1939). According to this view, spontaneous movements, originating endogenously within the CNS, are thought to be a very rare, indeed almost nonexistent phenomenon in embryos. Therefore, in spite of the obvious conceptual change that was occurring among students of the nervous system concerning the importance of *endogenously* produced and patterned neural activity (Bullock, 1961, 1962), it is nevertheless true that the original promulgation by Hamburger (1963) of the view that endogenously produced behavior may be a major and important feature of normal embryonic and fetal behavioral development was, initially at least, an unpalatable suggestion (e.g., Gottlieb and Kuo, 1965; Kuo, 1967; Schneirla, 1966). It is heartening to note that more recently Gottlieb (1973, p. 49) has fully recognized and acknowledged the validity of this view. Hamburger was led to consider this possibility by his rediscovery of the frequently neglected fact of motor primacy in the chick embryo, and by his observation that the regular periodic pattern of movements in the chick embryo appeared to be unrelated to any similarly periodic environmental fluctuation. It was obvious from the beginning, however, that these two facts alone were not sufficient to conclusively prove the existence of endogenous or spontaneous behavior. Most of the subsequent experiments that were done to prove or disprove this contention have relied upon sensory deprivation or deafferentation. It is to these studies that I now want to turn.

Perhaps the most sensitive and definitive test of the role of sensory input is the surgical deafferentation experiment. Although some caution must be exercised in the interpretation of the evidence from such studies (e.g., Oppenheim, 1972d), this technique, nevertheless, has the potential to determine which aspects of behavior are or are not dependent upon sensory input.

One of the first attempts to carry out such an experiment with the chick embryo involved partial deafferentation of the trunk and appendages (Hamburger *et al.*, 1966). This was accomplished by surgically removing the population of neuroblasts (neural crest cells) that later give rise to the spinal sensory ganglia in the lumbosacral spinal cord region, thereby depriving the legs and trunk in that region of *all* afferent innervation (see Fig. 8A). Furthermore, in order to eliminate the possibility that sensory input might be transmitted down the spinal cord from more rostral regions an entire segment of nervous tissue lying just in front of the deafferented region ("thoracic gap") was also removed. This resulted in a completely isolated neurobehavioral system consisting only of the lumbosacral spinal cord, the trunk, and the legs. The entire operative procedure was carried out at 2 days of incubation, prior to the onset of any function. Behavioral observation in this study was limited to the legs, and all operations were later verified histologically (Fig. 8C,D). A brief summary of the behavioral results is presented in Fig. 8B. This graph shows that at all of the ages examined the *amount* of cyclic motility, as compared to controls, was unimpaired by the deafferentation procedure; and, although not depicted here, the average *duration* of both *activity* and *inactivity* phases (periodicity) was also not modified. In general, the *character* of the movement patterns also did not differ from controls. Furthermore, the results of more recent deafferentation experiments with the chick have indicated that several important neurochemical substances in the spinal cord are also unmodified as a result of such an operation (Burt and Narayanan, 1970, 1972). In summary then, although attention in this study was only directed to certain select features of embryonic behavior, it can nevertheless be concluded that these features—which it should be remembered comprise a major component of the chick embryo's normal behavioral repertoire—can develop and be maintained without any assistance whatsoever from sensory stimulation. Since the only nerve cells remaining in the isolated spinal cord of the deafferented embryos consist of association cells or interneurons, and motoneurons (Fig. 8D), these results, when considered along with the electrophysiological data presented in Section IV,C, provide dramatic evidence for the existence of endogenously produced spontaneous behavior.

Other deafferentation experiments that have yielded behavioral results essentially similar to those reported above have involved bilateral removal of the trigeminal or fifth cranial ganglion which normally provides sensory innervation to the head, beak, and face (Hamburger and Narayanan, 1969),

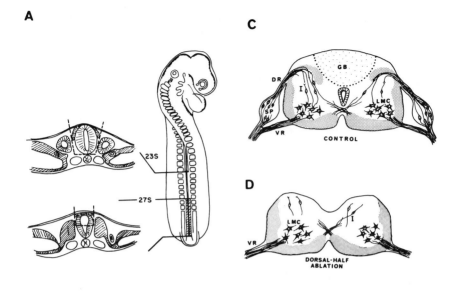

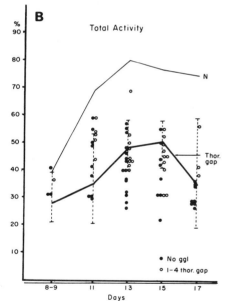

Fig. 8. Summary of dorsal-half ablation or spinal deafferentation experiment (see text). A, Schema of the operation showing removal of the neural crest in the lumbosacral region and removal of the entire spinal cord in the thoracic region. B, The behavioral results from this experiment. Total amount of cyclic activity (%) of unoperated (N), thoracic gap control, and deafferented embryos (circles). Solid circles indicate a complete absence of all sensory ganglia; open circles indicate residual ganglia in the thoracic region. Each circle represents the data for a single embryo. C, Cross section through the lumbosacral cord of a control embryo (GB, glycogen body; DR, dorsal root; SP, spinal ganglion; VR, ventral root; I, interneurons; LMC, lateral motor column). D, Cross section through the lumbosacral region of a deafferented embryo.

Note: The important comparison here is between the deafferented embryos (circles) and the *thoracic gap* control group; although both groups have thoracic gaps only the deafferented group lacks the spinal sensory system.

and bilateral removal of the entire inner ear, thereby depriving the embryo of both auditory and vestibular input (Decker, 1970).

In addition to these deafferentation studies there have been other kinds of deprivation experiments involving elimination of the amnion contractions (Oppenheim, 1966, 1972b); surgical amputation of the limbs, thereby depriving the embryo of a rich source of self-stimulation (Narayanan and Oppenheim, 1968; Helfenstein and Narayanan, 1969); modification of the position of the embryo in the egg (Oppenheim, 1970, 1973); and pharmacological immobilization (Provine, 1973). In all of these studies the deprived embryos differed little, if at all, from controls. The latter experiment is of particular interest because it involved the recording of the neural correlate of motility (i.e., bioelectric burst discharges) in twenty 15-day embryos *before* and *after* immobilization with curare. In other words, spinal cord bioelectric activity was recorded first while the embryo was overtly active and again after immobilization. There was not a significant difference between the mean percent of bioelectric burst activity during the two periods. Therefore, the movement-produced stimulation which was eliminated by curare would appear to have little, if any, immediate effect upon the neural correlate of motor activity in normal 15-day embryos.

While it is true that the results of all of the studies mentioned above are consistent in demonstrating a lack of effect of *sensory deprivation* upon certain features of the chick embryo's behavior, it would still be useful to know whether *sensory augmentation* might somehow modify this behavior, especially in light of the fact that the embryo can clearly respond to artificially applied stimulation after the seventh day of incubation. I have recently completed such a study in which chick embryos between the ages of 5 and 20 days of incubation were systematically subjected to greatly augmented levels of tactile and proprioceptive stimulation (Oppenheim, 1972d). In general, the results indicated that there were few, if any, differences in either cyclic motility or the actual frequency of movement between periods of augmented stimulation versus periods lacking such stimulation. I also found in this study that the responsivity of chick embryos to artificial tactile and proprioceptive stimuli tends to "habituate" rather rapidly; after the first few trials subsequent stimulation failed to elicit any further movements, although spontaneous responses continued unaffected. This finding may provide a partial answer to the seemingly contradictory situation in which the tactile and proprioceptive systems are known to be functional but where sensory deprivation or augmentation experiments consistently fail to demonstrate an effect on overt behavior. In other words, these results suggest that it may be advantageous for the chick embryo to "ignore," so to speak, much of the constant bombardment of sensory stimulation within the egg. By "ignore" I merely mean to stress the apparent lack of effect this stimulation has upon the *immediate, ongoing* behavior, and I certainly do not want to rule out the possibility that this stimulation may be potentially useful in the morphological or functional development of the nervous system, or for the manifestation of normal behavior *later* in development.

To summarize this section, it can be said that in practically every instance where an attempt has been made to systematically *analyze* the role of sensory input in the ongoing, overt behavior of the chick no support has been forthcoming for the contention that such input is a significant factor in the embryo's behavior. As I have attempted to indicate, however, such a conclusion is not tantamount to asserting that the chick embryo's behavior is completely refractory to sensory input, or that certain behavioral characteristics other than those stressed here might not be more dependent upon such input, or that later, even posthatching behavior, may not be impaired by early embryonic sensory restriction or augmentation. As was pointed out many years ago by Paul Weiss (1941), in what is one of the clearest and most insightful discussions ever presented on this problem, to argue that a functioning sensory system plays no role whatsoever in the development of motor behavior would be absurd! The relevant question is just what precisely is the role of such input? The experiments reviewed above represent the first attempt to answer this question for the chick embryo. As our techniques become more sophisticated and refined it would not be at all surprising to find that aspects of embryonic motor behavior, such as the timing and patterning of the activity of different muscles in the legs, are controlled, in part, by sensory stimulation. For example, it has recently been reported that monkeys, in which the arms and hands were deafferented during the perinatal period, fail to develop fine hand and arm coordination, although they can use the forelimbs during locomotion (Berman and Berman, 1973).

VI. Discussion and Conclusions

In the present essay I have attempted to present a comprehensive yet relatively concise summary of our current level of understanding of the ontogeny of behavior in the chick embryo. In doing so several important points have arisen which are substantially different from much past and present thought about behavioral development and therefore deserve further comment, particularly as they may be germane to problems that are of more direct concern to developmental psychologists.

The first, and perhaps most important assumption, underlying this entire essay has been that ultimately, in order to fully comprehend adult behavior, one will have to understand the ontogenetic precursors of that behavior, beginning with the very earliest embryonic movements. Although it is true that this idea has appeared sporadically in developmental psychology in the past (e.g., Kuo, 1922; Gesell, 1954; McGraw, 1946), Piaget is perhaps the only currently influential theorist who has consistently made this point. For instance, most recently he has stated that: " . . . the development of cognitive functions is a part of the epigenesis that leads from the first embryological stages to the adult state" (Piaget, 1970, p. 729). It is also my belief that problems of behavioral development can be most fruitfully attacked within the conceptual framework of developmental biology (i.e., an epigenetic and analytic approach). This view has frequently been expressed by Piaget (1970), as well as by several other investigators recently (Schneirla, 1966; Lenneberg, 1967; Eisenberg, 1971; Trevarthen, 1973). The rise of experimental embryology in this century is ample proof of the usefulness of such an approach.

The second general assumption that has guided many of the studies reviewed in this essay is that motor behavior, whether it be reflexogenous or endogenous in nature, *may be* an important aspect of behavioral development. By important I refer to the possibility that, in addition to its being simply a frequently occurring phenomenon, motor behavior during embryogenesis may be a necessary precursor for the ontogeny of later behavior. For example, the idea that "motility is the cradle of the mind" has been expressed in different ways by several authors (Sherrington, 1952; Coghill, 1929; Wolff, 1966; Herrick, 1956; Mittleman, 1960; Kestenberg, 1965), including Piaget, who states: "There is a continuous progression from spontaneous movements and reflexes to acquired habits *and from the latter to intelligence*" (Piaget and Inhelder, 1969, p. 5, italics added). Of course at the present time I realize that this suggestion represents no more than a working hypothesis which can only be proven or disproven by experiments done within an epigenetic framework. Furthermore, by making such a suggestion I do not want to imply that *all* behavioral processes may somehow be dependent upon motility, or that motility, as such, will ever provide a "window" to the mind. The available evidence certainly makes such simplistic notions rather untenable. I am well aware of the fact that for the most part, infant test scores, that primarily rely upon the attainment of various motor skills, are essentially unrelated to later cognition (Bayley, 1949, 1955). However, it might still be true that the normal acquisition of certain perceptual or "cognitive" skills may depend in part upon the prior orderly appearance and progression of specific motor patterns (e.g., see Wolff, 1966). I include here both spontaneous and stimulated movements, and the afferent feedback that might be a consequence of this movement. Indeed, such an assumption is explicit in Piaget's theory of human cognitive development. In this regard it is of interest, as White (1971) has pointed out, that many of the so-called "sensory" enrichment studies with human neonates and infants, have inadvertently produced increased motility along with an increased exposure to sensory stimuli. Therefore, it is not yet clear whether the positive effects found in such studies result from increased sensory stimulation, increased motility, or from some interaction of the two. Certainly the very nice experiments of Held and Hein (1963) with kittens provide clear evidence for the role of motor activity in the development of sensorimotor coordination.

The more specific question of whether *spontaneous* or *endogenously* produced motility as found in the chick embryo is also part of the normal behavior repertoire of higher forms, including man, and if so, whether it plays any role in later behavioral development, has received practically no attention at all (Hamburger, 1963). For instance, Hooker has admitted that in his own work on human fetuses, "The whole problem of spontaneous movements has been ignored" (Hooker, 1944, p. 36). Recently, however, Dawes *et al.* (1972), have elegantly demonstrated by deafferentation techniques that specific movements of the sheep fetus (respiratory-like) appear spontaneously throughout the latter two-thirds of the gestation period; and Narayanan *et al.* (1971) have provided evidence that spontaneous movements may also occur in the rat fetus. These recent reports, along with the well-established fact of spontaneous motility in the chick embryo, suggest that the investigation of spontaneous motility in mammalian fetuses might be a fruitful field for further inquiry.

In addition to the suggestion that embryonic and fetal motility may be a necessary precursor of later behavior, there is also the related possibility that this motility may be necessary for normal development and differentiation of the nervous system or other organ systems *during embryogenesis.* For example, it has now been well established that skeletal muscle contractions (i.e., motility) are essential during embryonic development in the chick for: (a) primary joint cavity formation; (b) the appearance of certain cartilages in the limbs; and (c) sculpturing of the form of the articular surfaces (Drachman and Sokoloff, 1966). It has also been reported that movements or motility may play a role in normal skeletal formation and muscular growth in fetal and neonatal mammals (Humphrey, 1971; Crawford, 1972; Walker, 1969). With regard to neurogenesis, Roffwarg *et al.* (1966) have suggested that the high levels of REM sleep characteristic of young mammals may be necessary to facilitate structural maturation and differentiation of key sensory and motor areas within the central nervous system. Both prenatal and neonatal motility could be the overt manifestation of uninhibited endogenous neural activity occurring during a REM-like sleep state. Indeed, in spite of the fact that there are serious problems in defining the various sleep states in immature organisms, some investigators have suggested that, during certain stages at least, embryonic motility might represent an overt sign of REM sleep (Corner *et al.,* 1973; Narayanan *et al.,* 1971; Wolff, 1966). Even if future studies would show no relationship between prenatal motility and REM sleep, however, the suggestion of Roffwarg *et al.* (1966) that endogenous neural activity may be necessary for certain aspects of neurogenesis deserves analytic attention. Of course the particular aspects or stages of neurogenesis that receive such attention should be dictated by an awareness of those neurogenetic events (such as the *de novo* formation of functional synapses, *in vitro*) that are already known not to require such neural activity (e.g., Crain *et al.,* 1968; Model *et al.,* 1971).

Finally there is the periodic or cyclic nature of the chick embryo's motility that is most striking in the early phases of incubation, but which also appears during the later prehatching and hatching period (see Section II,C). Although the possible significance of this rhythmicity for later behavior is still obscure, Wolff (1966) has suggested that cyclical motor activity (such as sucking) in the newborn mammal may be related to whatever mechanism controls the appropriate sequence and timing of motor behavior in adults, and consequently may be a necessary precursor for the appreciation of temporal order in the environment. In his important paper on serial order in behavior, Lashley (1951) also expressed the belief that rhythmic motor activities, " . . . form a sort of substratum upon which other activity is built. They contribute to every perception and to every integrated movement" (Lashley, 1951, p. 520).

In conclusion then, these new data from the chick embryo provide a fundamentally different frame of reference for understanding behavioral development. It is a frame of reference that has its roots in biology and embryology but which at the same time differs in several respects from the other general views that have arisen from these roots (i.e., Coghill and Windle, see discussion in Hamburger, 1963). This view should not in any way be regarded as a formal theory, however, but rather as a collection of observations and experiments that are suggestive of the direction in which some future investigations might proceed. When added to the already existing conceptions of functional development this new frame of reference should, at the very least, broaden the range of phenomena that are thought to be important in the ontogeny of embryonic behavior, and thereby hasten our understanding of this too often neglected period of development.

Acknowledgments

The writing of this essay was made possible by the continuing generous support of the North Carolina Department of Mental Health. The author's own research reported here was supported by grants from the National Science Foundation and the National Institute of Mental Health. Figures 8C and 8D were kindly drawn by Vernessa Riley. Ann Sterling provided competent editorial and typing assistance and Viktor Hamburger made several valuable suggestions on an earlier draft.

References

Alconero, B. B. 1965. The nature of the earliest spontaneous activity of the chick embryo. *J. Embryol. Exp. Morphol.* **13**, 255–266.

Ausubel, D., and Sullivan, E. V. 1970. "Theory and Problems of Child Development." Grune & Stratton, New York.

Bagust, J., Lewis, D. M., and Westerman, R. S. 1973. Polyneuronal innervation of kitten skeletal muscle. *J. Physiol.* **229**, 241–255.

Bangert, V. H. 1960. Untersuchungen zur Koordination der Kopf und Beinbewegungen beim Haushuhn. *Z. Tierpsychol.* **17**, 142–164.

Barcroft, J., and Barron, D. H. 1939. The development of behavior in foetal sheep. *J. Comp. Neurol.* **70**, 477–502.

Barker, J. L., and Nicoll, R. A. 1972. Gamma-aminobutyric acid: role in primary afferent depolarization. *Science* **176**, 1043–1045.

Barron, D. H. 1941. The functional development of some mammalian neuromuscular mechanisms. *Biol. Rev. Cambridge Phil. Soc.* **16**, 1–33.

Bayley, N. 1949. Consistency and variability in the growth of intelligence from birth to eighteen. *J. Genet. Psychol.* **75**, 165–196.

Bayley, N. 1955. On the growth of intelligence. *Amer. Psychol.* **10**, 805–818.

Berman, A. M., and Berman, D. 1973. Fetal deafferentation: the ontogenesis of movement in the absence of sensory feedback. *Exp. Neurol.* **38**, 170–176.

Bodian, D. 1968. Development of fine structure of spinal cord in monkey fetuses. II. Pre-reflex to period of long intersegmental reflexes. *J. Comp. Neurol.* **133**, 113–166.

Brinley, F. S. 1951. Effects of curare on fish embryos. *Physiol. Zool.* **24**, 186–195.

Bullock, T. H. 1961. The origins of patterned nervous discharge. *Behaviour* **17**, 48–59.

Bullock, T. H. 1962. Integration and rhythmicity in neural systems. *Amer. Zool.* **2**, 97–104.

Bursian, A. V. 1965. Primitive forms of photosensitivity at early stages of embryogenesis in the chick. *J. Evol. Biochem. Physiol.* **1**, 435–441. (Transl. from Russ.)

Bursian, A. V. 1969. Electromyographic study of rhythmic spontaneous activity in chick embryos. *J. Evol. Biochem. Physiol.* **5**, 501–506. (Transl. from Russ.)

Burt, A. M. 1968. Acetylcholinesterase and choline acetyltransferase activity in the developing chick spinal cord. *J. Exp. Zool.* **169**, 107–112.

Burt, A. M., and Narayanan, C. H. 1970. Effect of extrinsic neuronal connections on development of acetylcholinesterase and choline acetyltransferase activity in the ventral half of the chick spinal cord. *Exp. Neurol.* **29**, 201–210.

Burt, A. M., and Narayanan, C. H. 1972. Development of glucose-6-phosphate malate, and glutamate dehydrogenase activities in the ventral half of the chick spinal cord in the absence of extrinsic neuronal connections. *Exp. Neurol.* **34**, 342–353.

Carmichael, L. 1926. The development of behavior in vertebrates experimentally removed from the influence of external stimulation. *Psychol. Rev.* **33**, 51–58.

Carmichael, L. 1927. A further study of the development of behavior in vertebrates experimentally removed from the influence of external stimulation. *Psychol. Rev.* **34**, 34–47.

Carmichael, L. 1928. A further experimental study of the development of behavior. *Psychol. Rev.* **35**, 253–260.

Clark, E. L., and Clark, E. R. 1914. On the early pulsations of the posterior lymphhearts in chick embryos: Their relation to the body movements. *J. Exp. Zool.* **17**, 373–394.

Coghill, G. E. 1929. "Anatomy and the Problem of Behavior." Hafner, New York. (Reprint, 1964).)

Corner, M. A., and Bot, A. P. C. 1967. Developmental patterns in the central nervous system of birds. III. Somatic motility during the embryonic period and its relation to behavior after hatching. *Progr. Brain Res.* **26**, 214–236.

Corner, M. A., Schadé, J. P., Sedláček, J., Stoeckart, R., and Bot, A. P. C. 1967. Developmental patterns in the central nervous system of birds. I. Electrical activity in the cerebral hemisphere, optic lobe and cerebellum. *Progr. Brain Res.* **26**, 145–192.

Corner, M. A., Bakhuis, W. L., and van Wingerten, C. 1973. Sleep and wakefulness during early life in the domestic chicken, and their relationship to hatching and embryonic motility. *In* "Studies on the Development of Behavior and the Nervous System" (G. Gottlieb, ed.), Vol. 1, pp. 245–279. Academic Press, New York.

Crain, S. M. 1974. Tissue culture models of developing brain function. *In* "Studies on the Development of Behavior and the Nervous System" (G. Gottlieb, ed.), Vol. 2, pp. 69–114. Academic Press, New York.

Crain, S. M., Bornstein, M. B., and Peterson, E. R. 1968. Maturation of cultured embryonic CNS tissues during chronic exposure to agents which prevent bioelectric activity. *Brain Res.* **8**, 363–372.

Crawford, G. N. C. 1972. The effect of temporary limitation of movement on the longitudinal growth of voluntary muscle. *J. Anat.* **111**, 143–150.

Curtis, D. R., Duggan, A. W., and Johnston, G. A. R. 1969. Glycine, strychnine, picrotoxin and spinal inhibition. *Brain Res.* **14**, 759–762.

Dawes, G. S., Fox, H. E., Leduc, B. M., Liggins, G. D., and Richards, R. T. 1972. Respiratory movements and rapid eye movement sleep in the foetal lamb. *J. Physiol.* (*London*) **220**, 119–143.

Decker, J. D. 1967. Motility of the turtle embryo, *Chelydra serpentina. Science* **157**, 952–954.

Decker, J. D. 1970. The influence of early extirpation of the otocysts on development of behavior of the chick. *J. Exp. Zool.* **174**, 349–364.

Decker, J. D., and Hamburger, V. 1967. The influence of different brain regions on periodic motility in the chick embryo. *J. Exp. Zool.* **165**, 371–384.

Drachman, D. B. 1963. The developing motor end-plate: Pharmacological studies in the chick embryo. *J. Physiol.* (*London*) **169**, 707–712.

Drachman, D. B., and Sokoloff, L. 1966. The role of movement in embryonic joint formation. *Develop. Biol.* **14**, 401–420.

Eisenberg, L. 1971. Persistent problems in the study of the biopsychology of development. *In* "The Biopsychology of Development" (E. Tobach, L. R. Aronson, and E. Shaw, eds.), pp. 515–529. Academic Press, New York.

Foelix, R. F., and Oppenheim, R. W. 1973. Synaptogenesis in the avian embryo: Ultrastructure and possible behavioral correlates. *In* "Studies on the Development of Behavior and the Nervous System" (G. Gottlieb, ed.), Vol. 1, pp. 103–139. Academic Press, New York.

Fromme, A. 1941. An experimental study of the factors of maturation and practice in the behavioral development of the embryo of the frog, *Rana pipiens*. *Genet. Psychol. Monogr.* **24**, 219–256.

Gesell, A. L. 1954. The ontogenesis of infant behavior. *In* "Manual of Child Psychology" (L. Carmichael, ed.), pp. 335–373. Wiley, New York.

Giacobini, G., Marchisio, P. C., Giacobini, E., and Koslow, S. H. 1970. Developmental changes of cholinesterases and monoamine oxidase in chick embryo spinal and sympathetic ganglia. *J. Neurochem.* **17**, 1171–1185.

Gottlieb, G. 1968. Prenatal behavior of birds. *Quart. Rev. Biol.* **43**, 148–174.

Gottlieb, G. 1970. Conceptions of prenatal behavior. *In* "Development and Evolution of Behavior" (L. R. Aronson, E. Tobach, D. S. Lehrman, and J. S. Rosenblatt, eds.), pp. 111–137. Freeman, San Francisco, California.

Gottlieb, G. 1971. Ontogenesis of sensory function in birds and mammals. *In* "The Biopsychology of Development" (E. Tobach, L. R. Aronson, and E. Shaw, eds.), pp. 67–128. Academic Press, New York.

Gottlieb, G. 1973. Introduction to behavioral embryology. *In* "Behavioral Embryology, Studies on the Development of Behavior and the Nervous System," (G. Gottlieb, ed.) Vol. 1, pp. 3–50. Academic Press, New York.

Gottlieb, G., and Kuo, Z.-Y. 1965. Development of behavior in the duck embryo. *J. Comp. Physiol. Psychol.* **59**, 183–188.

Hamburger, V. 1960. "A Manual of Experimental Embryology." Univ. of Chicago Press, Chicago, Illinois.

Hamburger, V. 1963. Some aspects of the embryology of behavior. *Quart. Rev. Biol.* **38**, 342–365.

Hamburger, V. 1970. Embryonic motility in vertebrates. *In* "The Neurosciences, Second Study Program" (F. O. Schmitt, ed.), pp. 141–151. Rockefeller Univ. Press, New York.

Hamburger, V. 1971. Development of embryonic motility. *In* "The Biopsychology of Development" (E. Tobach, L. R. Aronson, and E. Shaw, eds.), pp. 45–66. Academic Press, New York.

Hamburger, V. 1973. Anatomical and physiological basis of embryonic motility in birds and mammals. *In* "Studies on the Development of Behavior and the Nervous System" (G. Gottlieb, ed.), Vol. 1, pp. 52–76. Academic Press, New York.

Hamburger, V., and Balaban, M. 1963. Observations and experiments on spontaneous rhythmical behavior in the chick embryo. *Develop. Biol.* **7**, 533–545.

Hamburger, V., and Levi-Montalcini, R. 1950. Some aspects of neuroembryology. *In* "Genetic Neurology" (P. Weiss, ed.), pp. 128–160. Univ. of Chicago Press, Chicago, Illinois.

Hamburger, V., and Narayanan, C. H. 1969. Effects of the deafferentation of the trigeminal area on the motility of the chick embryo. *J. Exp. Zool.* **170**, 411–426.

Hamburger, V., and Oppenheim, R. W. 1967. Prehatching motility and hatching behavior in the chick. *J. Exp. Zool.* **166**, 171–204.

Hamburger, V., Balaban, M., Oppenheim, R., and Wenger, E. 1965. Periodic motility of normal and spinal chick embryos between 8 and 17 days of incubation. *J. Exp. Zool.* **159**, 1–14.

Hamburger, V., Wenger, E., and Oppenheim, R. W. 1966. Motility in the chick embryo in the absence of sensory input. *J. Exp. Zool.* **162**, 133–160.

Harrison, R. G. 1904. An experimental study of the relation of the nervous system to the developing musculature in the embryo of the frog. *Amer. J. Anat.* **3**, 197–220.

Held, R., and Hein, A. 1963. Movement-produced stimulation in the development of visually guided behavior. *J. Comp. Physiol. Psychol.* **56**, 872–876.

Helfenstein, M., and Narayanan, C. H. 1969. Effects of bilateral limb-bud extirpation on motility and prehatching behavior in chicks. *J. Exp. Zool.* **172**, 233–244.

Herrick, C. J. 1956. "The Evolution of Human Nature." Harper, New York.

Hirano, H. 1967. Ultrastructural study on the morphogenesis of the neuromuscular junction in the skeletal muscle of the chick. *Z. Zellforsch. Mikrosk. Anat.* **79**, 198–208.

Hooker, D. 1944. "The Origin of Overt Behavior." Univ. of Michigan Press, Ann Arbor. (Univ. of Michigan Lecture, 1943.)

Hughes, A. 1966. Spontaneous movements in the embryo of *Eleutherodactylus martinicensis. Nature (London)* **211**, 51–53.

Hughes, A. 1968. Development of limb innervation. *In* "Growth of the Nervous System" (G. E. W. Wolstenholme and M. O'Connor, eds.), pp. 110–125. Little, Brown, Boston, Massachusetts.

Hughes, A., Bryant, A., and Bellairs, A. 1967. Embryonic behavior in the lizard, *Lacerta vivipara. J. Zool.* **153**, 139–152.

Humphrey, T. 1971. Development of oral and facial motor mechanisms in human fetuses and their relation to craniofacial growth. *J. Dent. Res.* **50**, 1428–1441.

Impekoven, M. and Gold, P. 1973. Prenatal origins of parent-young interactions in birds: a naturalistic approach. *In* "Studies on the Development of Behavior and the Nervous System" (G. Gottlieb, ed.), Vol. 1, pp. 326–356. Academic Press, New York.

Kaspar, J. L. 1964. The origin and development of motor patterns of the domestic chick. Ph.D. Thesis, Univ. of Wisconsin, Madison.

Kellogg, C., Vernadakis, A., and Rutledge, C. O. 1971. Uptake and metabolism of [^{3}H] norepinephrine in the cerebral hemispheres of chick embryos. *J. Neurochem.* **18**, 1931–1938.

Kestenberg, J. S. 1965. The role of movement patterns in development: I. Rhythms of movement. *Psychoanal. Quart.* **34**, 1–36.

Kovach, J. K. 1970. Development and mechanisms of behavior in the chick embryo during the last five days of incubation. *J. Comp. Physiol. Psychol.* **73**, 392–406.

Kuhlmann, F. 1909. Some preliminary observations on the development of instincts and habits in young birds. *Psychol. Rev., Monogr.* **1**, 49–84.

Kuo, Z.-Y. 1922. How are instincts acquired? *Psychol. Rev.* **29**, 344–365.

Kuo, Z.-Y. 1932a. Ontogeny of embryonic behavior in Aves. I. The chronology and general nature of the behavior of the chick embryo. *J. Exp. Zool.* **61**, 395–430.

Kuo, Z.-Y. 1932b. Ontogeny of embryonic behavior in Aves. II. The mechanical factors in the various stages leading to hatching. *J. Exp. Zool.* **62**, 453–483.

Kuo, Z.-Y. 1932c. Ontogeny of embryonic behavior in Aves. III. The structural and environmental factors in embryonic behavior. *J. Comp. Psychol.* **13**, 245–271.

Kuo, Z.-Y. 1932d. Ontogeny of embryonic behavior in Aves. IV. The influence of embryonic movements upon behavior after hatching. *J. Comp. Psychol.* **14**, 109–122.

Kuo, Z.-Y. 1932e. Ontogeny of embryonic behavior in Aves. V. The reflex concept in the light of embryonic behavior in birds. *Psychol. Rev.* **39**, 499–515.

Kuo, Z.-Y. 1939a. Studies in the physiology of the embryonic nervous system. I.

Effect of curare on motor activity of the chick embryo. *J. Exp. Zool.* **8**, 371–396.

Kuo, Z.-Y. 1939b. Studies in the physiology of the embryonic nervous system. IV. Development of acetylcholine in the chick embryo. *J. Neurophysiol.* **2**, 488–493.

Kuo, Z.-Y. 1967. "The Dynamics of Behavior Development." Random House, New York.

Lashley, K. S. 1951. The problem of serial order in behavior. *In* "Cerebral Mechanisms in Behavior" (L. A. Jeffress, ed.), pp. 112–136. Wiley, New York.

Lenneberg, E. H. 1967. "Biological Foundations of Language." Wiley, New York.

Levi-Montalcini, R., and Visintini, F. 1938a. Azione del curaro, della stricnina, dell'eserina, dell'acetilcolina sulla trasmissione dell'influsso nell'embrione di pollo dal 4 all'8 giorno d'incubazione. *Soc. Ital. Biol. Sper., Naples Boll.* **13**, 979–981.

Levi-Montalcini, R., and Visintini, F. 1938b. Eccitabilita e movimenti spontanei nell'embrione di pollo dol 4 all'8 giorno d'incubazione. *Soc. Ital. di Biol. Sper., Naples Boll.* **13**, 976–978.

Lewis, M. R. 1915. Rhythmical contraction of the skeletal muscle tissue observed in tissue cultures. *Amer. J. Physiol.* **38**, 153–161.

McGraw, M. B. 1939. Swimming behavior of the human infant. *J. Pediat.* **15**, 485–490.

McGraw, M. B. 1946. Maturation of behavior. *In* "Manual of Child Psychology" (L. Carmichael, ed.), pp. 332–369. Wiley, New York.

Matthews, S. A., and Detwiler, S. R. 1926. The reactions of *Amblystoma* embryos following prolonged treatment with chloretone. *J. Exp. Zool.* **45**, 279–292.

Mittleman, B. 1960. Intrauterine and early infantile motility. *Psychoanal. Study Child.* **15**, 104–127.

Model, R., Bornstein, M. B., Crain, S. M., and Pappas, G. D. 1971. An electron microscopic study of the development of synapses in cultured fetal mouse cerebrum continuously exposed to xylocaine. *J. Cell Biol.* **49**, 362–371.

Mumenthaler, E., and Engel, W. K. 1961. Cytological localization of cholinesterase in developing chick embryo skeletal muscle. *Acta Anat.* **47**, 274–299.

Narayanan, C. H. 1970. Apparatus and current techniques in the preparation of avian embryos for microsurgery and for observing embryonic behavior. *BioScience* **20**, 869–870.

Narayanan, C. H., and Oppenheim, R. W. 1968. Experimental studies on hatching behavior in the chick. II. Extirpation of the right wing. *J. Exp. Zool.* **168**, 395–402.

Narayanan, C. H., Fox, M. W., and Hamburger, V. 1971. Prenatal development of spontaneous and evoked activity in the rat (*Rattus norwegicus albinus*). *Behaviour* **40**, 100–134.

Nice, M. M. 1962. Development of behavior in precocial birds. *Trans. Linn. Soc.* **8**, 1–211.

Oppenheim, R. W. 1966. Amniotic contraction and embryonic motility in the chick embryo. *Science* **152**, 528–529.

Oppenheim, R. W. 1970. Some aspects of embryonic behavior in the duck (*Anas platyrhynchos*). *Anim. Behav.* **18**, 335–352.

Oppenheim, R. W. 1972a. Prehatching and hatching behavior in birds: A comparative study of altricial and precocial species. *Anim. Behav.* **20**, 644–655.

Oppenheim, R. W. 1972b. Embryology of behavior in birds: A critical review of the role of sensory stimulation in embryonic movement. *Proc. Int. Congr. Ornithol. 15th*, pp. 283–302.

Oppenheim, R. W. 1972c. Experimental studies on hatching behavior in the chick. III. The role of the midbrain and forebrain. *J. Comp. Neurol.* **146,** 479–506.

Oppenheim, R. W. 1972d. An experimental investigation of the possible role of tactile and proprioceptive stimulation in certain aspects of embryonic behavior in the chick. *Develop. Psychobiol.* **5,** 71–91.

Oppenheim, R. W. 1973. Prehatching and hatching behavior: A comparative and physiological consideration. *In* "Studies on the Development of Behavior and the Nervous System" (G. Gottlieb, ed.), Vol. 1, pp. 163–244. Academic Press, New York.

Oppenheim, R. W., and Foelix, R. F. 1972. Synaptogenesis in the chick embryo spinal cord. *Nature (London)* **235,** 126–128.

Oppenheim, R. W., and Narayanan, C. H. 1968. Experimental studies on hatching behavior in the chick. I. Thoracic spinal gaps. *J. Exp. Zool.* **168,** 387–394.

Oppenheim, R. W., Levin, H., and Harth, M. S. 1973. An investigation of various egg-opening techniques for use in avian behavioral embryology. *Develop. Psychobiol.* **6,** 53–68.

Orr, D. W., and Windle, W. F. 1934. The development of behavior in chick embryos: The appearance of somatic movements. *J. Comp. Neurol.* **60,** 271–285.

Patten, B. M. 1958. "Foundations of Embryology." McGraw-Hill, New York.

Piaget, J. 1970. Piaget's theory. *In* "Carmichael's Manual of Child Psychology" (P. H. Mussen, ed.), pp. 703–732. Wiley, New York.

Piaget, J., and Inhelder, B. 1969. "The Psychology of the Child." Routledge & Kegan, London.

Prestige, M. C. 1970. Differentiation, degeneration and the role of the periphery: Quantitative considerations. *In* "The Neurosciences, Second Study Program" (F. O. Schmitt, ed.), pp. 73–82. Rockefeller Univ. Press, New York.

Prestige, M. C., and Wilson, M. A. 1972. Loss of axons from ventral roots during development. *Brain Res.* **41,** 467–470.

Preyer, W. 1885. "Specielle Physiologie des Embryo." Grieben, Leipzig.

Provine, R. R. 1971. Embryonic spinal cord: Synchrony and spatial distribution of polyneuronal burst discharges. *Brain Res.* **29,** 155–158.

Provine, R. R. 1972. Ontogeny of bioelectric activity in the spinal cord of the chick embryo and its behavioral implications. *Brain Res.* **41,** 365–378.

Provine, R. R. 1973. Neurophysiological aspects of behavioral development in the chick embryo. *In* "Studies on the Development of Behavior and the Nervous System" (G. Gottlieb, ed.), Vol. 1, pp. 77–102. Academic Press, New York.

Provine, R. R., Sharma, S. C., Sandel, T. T., and Hamburger, V. 1970. Electrical activity in the spinal cord of the chick embryo *in situ*. *Proc. Nat. Acad. Sci. U.S.* **65,** 508–515.

Rhines, R. 1943. An experimental study of the development of the medial longitudinal fasciculus in the chick. *J. Comp. Neurol.* **79,** 107–126.

Ripley, K. L., and Provine, R. R. 1972. Neural correlates of embryonic motility in the chick. *Brain Res.* **45,** 127–134.

Roberts, A. 1971. The role of propagated skin impulses in the sensory system of young tadpoles. *Z. Vergl. Physiol.* **75,** 388–401.

Roberts, A., and Stirling, C. A. 1971. The properties and propagation of a cardiac-like impulse in the skin of young tadpoles. *Z. Vergl. Physiol.* **71,** 295–310.

Roffwarg, H. P., Muzio, J. N., and Dement, W. C. 1966. Ontogenetic development of the human sleep-dream cycle. *Science* **152,** 604–619.

Rogers, K. T., DeVries, L., Kepler, J. A., Kepler, C. R., and Speidel, E. R. 1960. Studies on chick brain of biochemical differentiation related to morphological differentiation and onset of function. II. Alkaline phosphatase and cholinesterase levels, and onset of function. *J. Exp. Zool.* **144**, 89–103.

Schneirla, T. C. 1966. Behavioral development and comparative psychology. *Quart. Rev. Biol.* **41**, 283–302.

Schooland, J. B. 1942. Are there any innate behavior tendencies? *Genet. Psychol. Monogr.* **25**, 219–287.

Sherrington, C. S. 1952. "Man, on His Nature." Cambridge Univ. Press, London and New York.

Simon, M. E. 1954. Der optomotorische Nystagmus während der Entwicklung normaler und optisch isoliert aufgewachsener Kücken. *Z. Vergleich. Physiol.* **37**, 82–105.

Spaulding, D. A. 1872. On instinct. *Nature (London)* **6**, 485–486.

Sviderskaya, G. E. 1967. Effect of sound on the motor activity of chick embryos. *Bull. Exp. Biol. Med. (USSR)* **63**, 24–28.

Tello, J. F. 1922. Die Entstehung der motorischen und sensiblen Nervenendigungen. I. In dem lokomotorischen system der höheren Wirbeltiere. Muskulate histogenese. *Z. Gesamte Anat., Abt. 1* **64**, 248–440.

Tracy, H. C. 1926. The development of motility and behavior reactions in the toadfish (*Opsanus tau*). *J. Comp. Neurol.* **40**, 253–369.

Tremor, J. W., and Rogallo, V. L. 1970. A small animal actoballistocardiograph: Description and illustrations for its use. *Physiol. Behav.* **5**, 247–251.

Trevarthen, C. B. 1973. Behavioral embryology. *In* "The Handbook of Perception" (E. C. Carterette and M. P. Friedman, eds.). Academic Press, New York.

Tuge, H. 1937. The development of behavior in avian embryos. *J. Comp. Neurol.* **66**, 157–179.

Vernadakis, A. 1969. Sensitivity of chick embryos to chlorpromazine. *Brain Res.* **12**, 223–226.

Vince, M. A. 1973. Some environmental effects on the activity and development of the avian embryo. *In* "Studies on the Development of Behavior and the Nervous System" (G. Gottlieb, ed.), Vol. 1, pp. 286–323. Academic Press, New York.

Visintini, F., and Levi-Montalcini, R. 1939. Relazione tra differenziazione strutturale e funzionale dei centri e delle vie nervose nell'embrione di pollo. *Schweiz. Arch. Neurol. Psychiat.* **43**, 1–45.

Walker, B. E. 1969. Correlation of embryonic movement with palate closure in mice. *Teratology* **2**, 191–198.

Weiss, P. 1939. "Principles of Development." Holt, New York.

Weiss, P. 1941. Self-differentiation of the basic patterns of coordination. *Comp. Psychol. Monogr.* **17**, 1–96.

Wenger, B. S. 1968. Construction and use of the vibrating needle for embryonic operations. *BioScience* **18**, 226–228.

White, B. L. 1971. "Human Infants, Experience and Psychological Development." Prentice-Hall, Englewood Cliffs, New Jersey.

Windle, W. F. 1950. Reflexes of mammalian embryos and fetuses. *In* "Genetic Neurology" (P. Weiss, ed.), pp. 214–222. Univ. of Chicago Press, Chicago, Illinois.

Wintrebert, M. P. 1930. La Contraction rhythmée des myotomes chez les embryons de selaciens: I. Observation de *Scylliorhinus canicula. Arch. Zool. Exp.* **60,** 221.

Wolff, P. H. 1966. The causes, controls and organization of behavior in the neonate. *Psychol. Issues Monogr. Ser.* **5** (1), Monogr. 17.

NOTE ADDED IN PROOF

In a recent article Condon and Sander (1974) (*Science* **183,** 99–101) have demonstrated a rather complex and subtle causal interaction between the sequence and rhythm of body motility in human neonates and the organized speech behavior of adults in the environment. That is, the infant's motor activity seems to be temporally organized by the vocal rhythms it hears. The authors suggest that this interaction may serve to help the infant acquire and organize certain aspects of its cultural language system long before it actually begins to speak the language. It may not be overly optimistic to think that the overt motility of vertebrate embryos will also eventually be shown to be an integral factor in the development of equally complex behavior patterns.

Processes Governing Behavioral States of Readiness

WALTER HEILIGENBERG[1]

MAX-PLANCK-INSTITUT FÜR VERHALTENSPHYSIOLOGIE
SEEWIESEN, GERMANY

I. INTRODUCTION

Certain behavioral patterns of animals are as typical for their species as are the physical features used by taxonomists. Whereas some of these patterns, like chirping in crickets, occur without any known external cause, others, like attacking in cichlid fish, occur in response to specific stimulus patterns, though not in a deterministic reflex-like manner. In either case, the probability that an animal will perform a particular behavioral pattern depends on environmental conditions, such as temperature, light, and time of day or year. Moreover, even under constant external conditions this probability may fluctuate considerably, though usually in some systematic way. In addition, some stimulus patterns are known to affect the probability of occurrence of certain behaviors over varying periods following their presentation. This paper is concerned with how the ideas of "time series analysis" and the exploitation of certain fortunate behavioral situations can furnish information about these phenomena.

[1] Present address: Scripps Institution of Oceanography, University of California at San Diego, La Jolla, California.

II. Methodological Considerations and Experimental Procedures

The present paper will concentrate on behavioral patterns which are fairly stereotyped and which vary little in intensity and duration. Thus, in a first approach, they can be dealt with as identical events occuring at times $t_1, t_2, \ldots, t_n$ within a given period of time T. Counting the number f of occurrences of a particular behavior within successive periods of time of standard length $T, T_1, T_2, \ldots, T_N$, one obtains a sequence of numbers $f_1, f_2, \ldots, f_N$ which represents the "rate" of occurrence of this behavior as a function of time.

The term "readiness" will be used here to describe the propensity of an animal to perform a particular behavior, and the state of readiness will be measured by the "rate" of occurrence of this behavior in a standardized environmental situation. If the behavior studied ordinarily occurs in response to certain stimulus patterns, a standard test stimulus has to be presented and the number of responses obtained per unit time taken as a measure of the animal's readiness to respond. However, in order to measure this readiness as a function of time, the test stimulus should be applied continuously, or at least continually at short intervals. This procedure, however, encounters the following problem.

When a particular stimulus is presented repeatedly and at sufficiently short intervals the animal "habituates." Whereas the animal may respond strongly to the first few presentations, subsequent presentations release weaker and weaker responses until a stationary level of habituation is reached which, in extreme cases, is so high that the animal ceases to respond at all. If the stimulus is not presented for a sufficiently long period of time thereafter recovery occurs, i.e., the animal responds strongly again if the stimulus is presented anew. If the animal has habituated to a given stimulus and a different stimulus releasing the same type of response is presented instead, the animal first responds strongly to the new stimulus but then habituates to this stimulus in the same way as to the first one. Processes of habituation have been studied in different animal species and a broad spectrum of different time courses of habituation and dishabituation found (cf. Hinde, 1970).

It appears from the above description that the level of habituation to a given stimulus does not reflect the general readiness of the animal to perform the associated response. Habituation appears instead to be a mechanism neutralizing the effect of stimuli which appear too frequently and thus convey little "information." For this reason it seems justifiable to consider the phenomenon of habituation as a process independent of the state of readiness. In order to measure the state of readiness by the rate of response at two different periods of time in a comparable manner the level of habituation to the test stimulus must therefore be the same. This requires a test stimulus which still releases responses at a sufficient rate when habituation has reached a steady level. A second condition a test stimulus should meet, of course, is that it affects the state of readiness to be measured as little as possible.

The behavioral system of an animal may be looked upon as a set of different states of readiness determining the probabilities of occurrence of corresponding behavioral outputs. In order to investigate intrinsic fluctuations of different states of readiness and mutual correlations between them, the system should first be studied under steady conditions, i.e., by keeping all known stimulus inputs constant. Subsequently, the effects of certain stimulus presentations on different states of readiness should be analyzed. This is achieved by presenting a stimulus for a short period of time without affecting the standardized situations in which the animal is kept, so that a particular state of readiness can be measured in the pre- and poststimulatory period in a comparable manner. Any systematic changes can then be attributed directly to the stimulus presented.

In the following sections examples will be given, showing fluctuations in a state of readiness under steady conditions (Section III) and different effects of stimulus presentations (Section IV). These examples deal with chirping in crickets and attacking in cichlid fish, two behavioral systems extensively studied by the author. In spite of the notable taxonomic differences between the species involved, the processes governing behavioral states of readiness are strikingly similar.

The following experimental procedures were applied.

The "spontaneous" chirping activity of single isolated crickets (*Acheta domesticus*) was recorded on tape and then analyzed on a digital computer. The acoustic responses of a cricket to another cricket's chriping were studied by playing certain chirp patterns from tape loops. These experiments were generally performed automatically, so that large amounts of data could be obtained.

In order to measure the attack readiness of a cichlid fish as a function of time, a suitable test stimulus had to be presented continually. As outlined above, this test stimulus had still to trigger attack responses at a sufficient rate after the animal had reached a stationary level of habituation. This was accomplished as follows.

A territorial male cichlid of the genus *Pelmatochromis* or *Haplochromis* attacks other fish which enter his territory. If the intruder is not too big, the male will rush at it and bite. A large intruder may be approached hesitantly: after mutually displaying their brightly colored sides and median fins (cf. Seitz, 1940, 1942; Baerends and Baerends-von Roon, 1950) the two fish engage in a fight unless the intruder interrupts its display and flees.

A fight may last several minutes, until one of the opponents turns away and escapes from further attacks by the victor.

If adult and juvenile fish of any cichlid species are placed in an aquarium occupied by a territorial male, both will be attacked at fluctuating rates over several weeks. If the rate of attacks directed at juveniles is compared with the rate of attacks directed at adults, a strong positive correlation $(r = 0.95)$ is found (Leong, 1969, Fig. 4). Therefore either rate can be taken as a measure of the attack readiness in the territorial male. For two reasons, however, juvenile fish provide a better test stimulus than adults. First, unlike adult opponents, juveniles never engage the territorial male in fights which may lead to his defeat and temporary loss of aggressive tendencies, and second, as will be shown in Section IV,C, juveniles affect the attack readiness in a territorial male very little.

In placing a given number of small fish in an aquarium occupied by a territorial male, one should take care that no niches are available for the small fish to hide in, since the rate of attacks directed at small fish depends on the number of small fish available.

In order to measure the effect of certain visual stimulus presentations on the attack readiness of a territorial male, the small fish which are to serve as a constant test stimulus were prevented from seeing the stimuli presented to the male so that their behavior would not be affected in advance (for further details of the experimental procedure, see Heiligenberg, 1965; Leong, 1969).

The experimental animals were observed from a hide and their behavior recorded on punch tape. Visual stimuli were presented by remote control.

III. Fluctuations in a State of Readiness under Steady Conditions

If a male cricket (*Acheta domesticus*) is isolated in a quiet room under constant environmental conditions its chirp rate fluctuates considerably with time. However, when the numbers f_k of chirps counted within successive 1-minute intervals $(k = 1, 2, \ldots)$ are compared, values close to one another in the sequence of data appear similar.

If a male cichlid fish (*Pelmatrochromis kribensis*) is placed in an aquarium together with a number of small test fish, the rate of attacks directed at these small fish fluctuates strongly from minute to minute without any apparent external cause. The rate of attacking seems to be less predictable than the rate of chirping in the cricket mentioned above.

Serial correlation analysis is a favorite mathematical tool for investigating fluctuations in an ordered sequence of numbers $f_1, f_2, \ldots, f_n$. The

attacking cichlid

chirping cricket

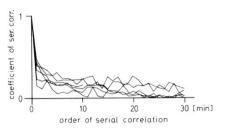
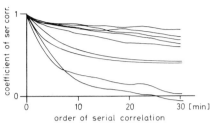

FIG. 1. Serial correlation coefficients plotted for attacking cichlid and chirping cricket. Different records of several animals were analyzed, the correlation coefficients for each record connected by continuous lines. Since larger sequences of data were available for chirping than for attacking, the curves in the right-hand diagram are less ragged. (From Heiligenberg, 1969.)

correlation coefficient of order k, written r_k, indicates the degree of similarity which is found between f-values separated by k locations in the sequence.[2] Maximum similarity, which necessarily applies for $k = 0$, results in a coefficient of $+1$. If no systematic relation is found between the f-values compared, the coefficient is zero. If f-values tend to be of opposite size, a given one being small when the other one is large, the correlation coefficient approaches -1. Performing a serial correlation analysis on the number of behavioral events counted in successive 1-minute intervals, one obtains the curves shown in Fig. 1. As one would expect from the qualitative description given above, the coefficient of correlation of order k, r_k, approaches zero at much larger k values in the case of chirping than in the case of attacking, which means that the rate of chirping can be "predicted ahead" more accurately over a longer period of time than the rate of attacking.

At this point one could argue that the lesser predictability of the rate of attacking was due to the "randomness" of the behavior of the small test fish moving around in the aquarium. However, a similarly low predictability was also found in the occurrence of a food-searching pattern, which consisted in sifting small amounts of substrate. This behavior was recorded in totally

[2] Given a sequence of numbers $f_1, f_2, \ldots, f_N$, the serial correlation coefficient of order k is

$$r_k = \frac{\sum_{i=1}^{N-k} f_i \cdot f_{i+k} - \frac{1}{N-k} \cdot \sum_{i=1}^{N-k} f_i \cdot \sum_{i=1}^{N-k} f_{i+k}}{[\sigma_{(1,N-k)} \cdot \sigma_{(k+1,N)}]^{1/2}}$$

with

$$\sigma_{(n,m)} = \sum_{i=n}^{m} f_i^2 - \frac{1}{m-n+1} \left(\sum_{i=n}^{m} f_i \right)^2$$

isolated adult cichlids in tanks with a homogeneous sand substrate free of food particles. The random behavior of the test fish therefore is not necessarily the sole cause of the low predictability in attack rate.

Another way to demonstrate a relation between the number f of behavioral events counted in a given interval n and a subsequent interval $n + k$, is to plot the number of events f_{n+k}, i.e., the number of occurrences in the interval $n + k$, as a function of the number of events f_n. When this procedure is applied to a particular sequence of data $f_1, f_2, \ldots$ representing the number of chirps counted in successive 1-minute intervals, the diagram in Fig. 2a is obtained which shows a linear relation between f_n and f_{n+1}.

To demonstrate this linear relation further, f_n values were grouped within different ranges on the abscissa and the average f_n and f_{n+k} calculated for each range. These values were plotted in Fig. 2b for three different values of k. The corresponding regression lines were calculated from raw data as in Fig. 2a and follow the formula

$$\overline{f_{n+k}} = a_k \cdot f_n + c_k \tag{1}$$

Thus, the average rate of events to be expected in the kth interval following a given interval, is a linear function of the rate of events counted in that interval.

Since the set of all f_n-values and the set of all f_{n+k}-values are both drawn from the same data sample, their mean values and variances must be identical. For this reason the linear regression coefficient a_k is identical to the

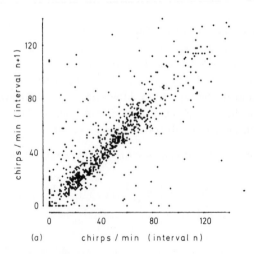

(a)   chirps / min (interval n)

Fig. 2a. The number of chirps counted in a given 1-minute period (ordinate) plotted versus the number counted in the previous period (abscissa). All points in the diagram were taken from one continuous record.

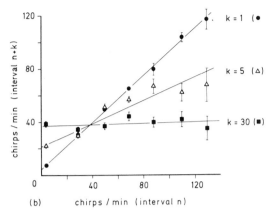

(b)

FIG. 2b. The average chirp rate found k minutes later plotted as a function of the present chirp rate. For three different values of k, raw data as shown in Fig. 2a) were grouped according to different ranges on the abscissa. Within each range, average values and variances of the mean were calculated with respect to both coordinates and plotted in Fig. 2b). Since average values on the abscissa were calculated from limited data ranges, the corresponding variances were too small to be plotted. The regression lines were calculated from raw data as shown in Fig. 2a).

serial correlation coefficient r_k.[3] Further, by taking the expected value, estimated by the empirical mean of both sides of Eq. (1) and denoting the average f-value by $\bar{f}$, it follows that

$$\bar{f} = a_k \cdot \bar{f} + c_k, \qquad \text{i.e.,} \qquad c_k = \bar{f} \cdot (1 - a_k) \tag{2}$$

The same procedure was also applied to sequences of data of attacking and sifting behavior recorded from male cichlids, and the same linear relations found, the only difference being that, since the serial correlation coefficient approaches zero more quickly, the corresponding lines of linear regression are less steep.

IV. The Effect of Stimulus Presentations on a State of Readiness

A. Presentations of a Single Stimulus Pattern

A territorial male cichlid may attack other fish in his environment at a very low rate. However, if another territorial male appears at the boundary

---

[3] Given a sequence of numbers f_1, f_2, . . . , f_N, the linear regression coefficient a_k is obtained by replacing $\sigma_{(k+1,N)}$ by $\sigma_{(1,N-k)}$ in the formula given for r_k (see footnote 2). These two expressions reach equal values for large values of N.

of his territory, his attack rate will increase immediately, even if the appearance of the rival was for so short an interval that a mutual display or a fight could not develop. It seems that the appearance of another male raises the readiness to attack in the resident male. In order to study this phenomenon quantitatively the following experiment was designed (Heiligenberg, 1965).

A territorial cichlid male was placed in an aquarium together with a certain number of small test fish to measure its readiness to attack. After a few days the adult male "settled down" in the new environment and attacked the small test fish at a moderate rate. Experiments were then performed as follows.

To measure the momentary baseline readiness to attack, the male was observed for 15 minutes, and then a dummy resembling a conspecific male was presented behind a glass partition for 30 seconds. The male usually stared at the dummy and did not engage in any further activity. Following this dummy presentation the male was observed for a further 30 minutes. As one can see in Fig. 3a, the attack rate of the male was raised by the presentation of the dummy and returned to its prestimulatory level with a half-time of approximately 1.5 minutes. No such increment in attack rate was found when a neutral object, such as a uniformly colored ball, was presented instead. Since the behavior of the small test fish cannot be affected by the visual stimulus presented only to the male, the increment in attack rate must be a measure of the animal's altered state of readiness to attack as a consequence of the dummy presentation.

In order to determine whether the size of the increment in attack rate depends on its prestimulatory level, the poststimulatory attack rate was plotted as a function of prestimulatory attack rate in Fig. 3b. As one can see in this diagram, the average attack rate found after the dummy presentation was higher by a constant amount than that expected with no dummy presentation, whatever its prestimulatory level.

If the stimulus presentation had raised the attack rate for only a brief period of time, such as the first poststimulatory minute, and did not have any further aftereffects, the average attack rate should return to its prestimulatory level in the same manner as the serial correlation function (cf. Fig. 1, left-hand diagram) approaches zero.[4] However, since the curves shown in Fig. 1 are clearly distinct from the fairly exponentially shaped return

---

[4] This is shown as follows: Let m be the current number of the first poststimulatory 1-minute interval and f_m the attack rate counted. Then due to Eq. (1), the average attack rate to be expected k intervals later is

$$\overline{f_{m+k}} = a_k \cdot f_m + c_k$$

with $c_k = \bar{f} \cdot (1 - a_k)$, $\bar{f}$ being the general rate level, as represented in the prestimulatory

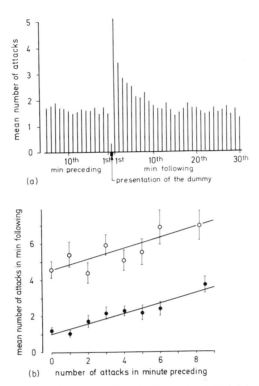

(a)

(b)

FIG. 3a. The average attack rate (ordinate) of a male cichlid fish (*Pelmatochromis kribensis*) before and after the presentation of a dummy resembling a conspecific male. After a quick rise the attack rate returns exponentially to its prestimulatory level with a half-time of approximately 1.5 minutes. Data of 150 experiments were averaged. The half-time is defined as the period that it takes for an exponential process to approach its asymptote by 50% from its initial deviation.

FIG. 3b. The average number of attacks counted in the poststimulatory minute plotted as a function of the number of attacks in the prestimulatory minute. The average poststimulatory attack rate is by a constant amount higher (open circles) than *expected* in the case of no dummy presentation (filled circles). Variances of the mean are indicated by vertical bars. Prestimulatory attack rates larger than six were rare. For this reason they were pooled and their average plotted on the abscissa. (From Heiligenberg, 1965.)

of the attack rate to its prestimulatory level as shown in Fig. 3a, this hypothesis must be rejected. One must therefore assume that the presentation of

---

part of the diagram in Fig. 3a, this is equivalent to

$$\overline{f_{m+k}} = a_k \cdot (f_m - \bar{f}) + \bar{f}$$

Since $a_k = r_k$ (as shown in footnote 3), $\overline{f_{m+k}} - \bar{f}$ is proportional to r_k for any given f_m, i.e., $\overline{f_{m+k}} - \bar{f}$ approaches zero in the same manner as r_k, $k = 1, 2, \ldots$.

the dummy causes an excitatory process reaching its peak within the first poststimulatory minute and then decaying with a half-time of approximately 1.5 minutes. This process is additively superimposed upon the level of attack readiness expected without stimulation. With A being the additive increment reached within the first poststimulatory minute $(t = 0)$, and α being the rate constant corresponding to a half-time of 1.5 minutes, this process $E(t)$ can be written

$$E(t) = A \cdot \exp\, (-\alpha \cdot t),\, t \geq 0 \tag{3}$$

$\exp\,(x)$ denoting the exponential function of x.

This formula represents a continuous approximation to the poststimulatory part of the histogram shown in Fig. 3a, to the extent that it exceeds the level expected without stimulation.

Other processes of this nature will be described in the following sections.

B.   Combined Presentations of Different Stimulus Patterns

Applying the method outlined in Section IV,A, Leong (1969) presented fish dummies showing different color patterns of the cichlid species *Haplochromis burtoni* and determined their effects on the attack readiness in adult males of this species. The aim of this investigation was first to measure the effect of single color markings and then to determine how different combinations of such markings would affect the readiness to attack. Among all the color markings tested, only two were found to affect a male's readiness to attack: a black vertical eye-bar and a field of orange spots above the pectoral fins. The black eye-bar caused an additive increment in attack rate reaching its peak somewhat more slowly than described in Section A for *Pelmatochromis kribensis,* and vanishing with a half-time of approximately 3 minutes. The field of orange spots caused a decrement in attack rate which vanished with a half-time of the order of 10 minutes. When both color patterns were present on the dummy, the total effect was the sum of the effects caused by each color pattern when presented alone (Fig. 4).

Since the number of experiments performed with dummies showing only the field of orange spots was not large, the nature of the "inhibitory" process caused by this color pattern has not yet been analyzed in detail. However, fom the data available so far, one can not rule out that the rate of attacking is decreased by an amount which is independent of the prestimulatory level of attack rate.

Since the curves given in Fig. 4 represent average values, one could argue that if a dummy with both color patterns is presented, the fish may respond in 50% of all experiments only to the black eye-bar and in all other experiments only to the field of orange spots and that the shape of the middle

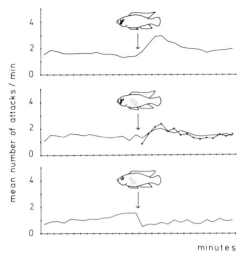

mean number of attacks / min

minutes

FIG. 4. Changes in average attack rate (ordinates) of a male *Haplochromis burtoni* caused by the presentation (arrows) of three different dummies. The attack rate is rasied, when a dummy with a black vertical eye-bar is presented (upper curve, average of 220 experiments), and lowered, when a dummy with a field of orange spots above the pectoral fins is shown (lower curve, average of 90 experiments). An intermediate effect is obtained, when a dummy combining both colorations is presented (middle curve, average of 150 experiments). The curve marked by asterisks represents the expected time course when the effects on attack rate as shown in the upper and lower curves of the figure were added. Since a particular dummy was often presented at half-hour intervals some aftereffect is still found in the prestimulatory 15-minute period. (All curves calculated from raw data of Leong, 1969.)

curve in Fig. 4 is an artifact due to averaging data of *all* experiments. This argument does not hold, however. When the increment in attack rate is calculated for each experiment, and the relative frequencies of different increment values found for each type of dummy are plotted, estimated probability distributions are found which differ only in their mean values (Fig. 5). According to the argument stated above, however, one would expect the probability distribution for the dummy combining both color patterns to show either two peaks or at least a larger variance. One must therefore assume that if both color patterns are presented on a dummy, each pattern has its own individual effect and that the two effects superimpose additively.

As Leong (1969) found, the black vertical eye-bar in *Haplochromis burtoni* is the only component of a more complex color pattern of the head affecting the readiness to attack in other males. A recent investigation (Heiligenberg *et al.*, 1972), using dummies with eye-bars that could be rotated around the center of the eye, showed that the angular orientation with re-

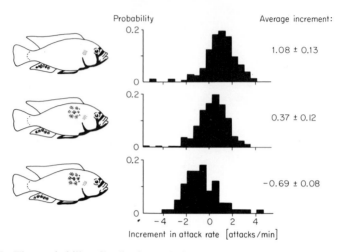

FIG. 5. The probability distributions of the increment in attack rates in a male *Haplochromis burtoni* caused by three different dummies. The distributions differ only in their mean value written on the right. The dummies are drawn in more detail than in Fig. 4 and are characterized from top to bottom as follows: vertical black eye-bar and no orange spots, vertical black eye-bar and orange spots, no vertical black eye-bar and orange spots. (From Leong, 1969, scale on the abscissa altered.[5])

spect to the body coordinates of the dummy is one of the most relevant stimulus parameters affecting the readiness to attack. In order to determine whether a fish perceives this parameter in a manner independent of the posture of the dummy, dummies were presented horizontally as well as vertically with their heads pointing downward. This head-down posture is assumed by cichlids while feeding and digging on the ground and most conspicuously in a so-called "threat display" which is likely to occur when, for instance, a parental fish guarding its offspring hesitates to attack an intruder.

The increments in attack rate, caused by five different angular eye-bar orientations in horizontally and vertically presented dummies are shown in Fig. 6. This diagram shows that the increment in attack rate is higher the more the eye-bar parallels the profile of the forehead, no matter in which posture the dummy is presented. However, the head-down posture itself is associated with a further increment in attack rate superimposed upon the increment due to the black eye-bar. Since the variances involved are very large and the effect of the head-down posture rather small, one cannot decide whether this superposition is additive or multiplicative, i.e., whether

[5] In the present paper "attack rate" is measured by the number of events per unit time. Leong measured attack rate by the reciprocal of the average interevent interval. The difference in these procedures is discussed elsewhere (Heiligenberg *et al.*, 1972).

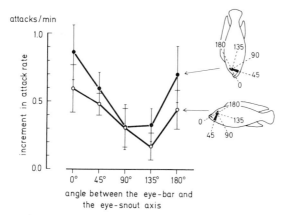

FIG. 6. The increment in attack rate in a male *Haplochromis burtoni* caused by a dummy as a function of eye-bar orientation and dummy posture. The increment is higher the more the black eye-bar parallels the profile of the forehead. A dummy presented head down is more effective than a dummy presented horizontally. The natural eye-bar orientation is 45°. Variances of the mean are indicated by vertical bars. (From Heiligenberg *et al.*, 1972.)

the head-down posture contributes a constant increment added to the effect of any particular eye-bar orientation, or whether it "amplifies" the effect of the eye-bar by a factor of the order of 1.5. Combining the stimulus components "head down" and "field of orange spots above pectorals" should help to decide between these alternatives, since then an additive effect would increase the attack rate and a multiplicative one might well reduce it.

C. REPEATED PRESENTATIONS OF A STIMULUS PATTERN: SHORT- AND LONG-TERM EFFECTS

When a single stimulus chirp is played to a chirping cricket, the cricket will briefly raise its chirp rate provided the stimulus does not coincide with one of its own chirps, in which case it has no effect. After the chirp rate has risen it returns to its prestimulatory level with a half-time of approximately 2 seconds (Fig. 7a). Plotting the poststimulatory chirp rate as a function of prestimulatory chirp rate, it is found that the stimulus chirp causes an additive increment in chirp rate (Fig. 7b). This process is thus of the same nature as the phenomenon discussed in Section A, Fig. 3, the only difference being that the time constant is of the order of seconds rather than minutes.

When single stimulus chirps are presented at 10-second intervals over a period of 3 minutes, each single stimulus causes the effect described above. However, after the last stimulus has been presented, the chirp rate first falls with a half-time of 2 seconds, but then, with a much longer time constant,

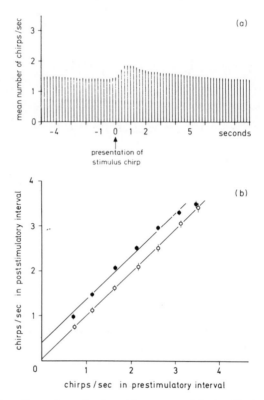

Fig. 7a. The effect of a stimulus chirp on a cricket's chirping. The chirp rate is raised in response to a single stimulus chirp and returns to its prestimulatory level with a half-time of approximately 2 seconds. Average data of 4000 experiments are shown.

Fig. 7b. Poststimulatory chirp rate (filled circles) as a function of prestimulatory (controls indicated by open circles). The increment in chirp rate, as represented by the vertical distance of the parallel regression lines, does not depend on the prestimulatory chirp rate level below highest possible values. Intervals of 2.5 seconds preceding and following a single stimulus chirp were compared, the chirp rate calculated as the reciprocal of the average interchirp interval. Prestimulatory chirp rates were grouped according to different ranges on the abscissa; mean values and variances of the mean were calculated for each group with respect to both coordinates. (From Heiligenberg, 1969.)

returns to the level measured before the first stimulus was presented (Fig. 8). This leads to the assumption that each single stimulus, apart from causing an additive increment S_1 which decays again with a half-time of 2 seconds, also causes a comparatively smaller additional increment S_2 which decays with a half-time of approximately 140 seconds, as estimated from the slow decay in Fig. 8. Whereas the short-term process, characterized by a half-time of 2 seconds, has decayed nearly completely when the subse-

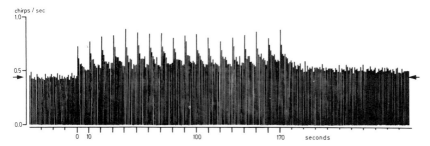

FIG. 8. The effect of repeated stimulus chirps on a cricket's chirping. As 18 stimulus chirps are presented at 10-second intervals (extended markings on the abscissa) the chirp rate rises slowly and stays above the prestimulatory level (indicated by horizontal arrows) for many seconds. Average data of 500 experiments with prestimulatory chirp rate levels between 0 and 1 chirps per second are shown. (From Heiligenberg, 1969.)

quent stimulus is presented, the long-term process, characterized by a half-time of 140 seconds, is still on a high level and thus builds up stepwise if stimuli are presented in succession.

The total excitatory effect $E(t)$ caused by a single stimulus chirp can thus be written as the sum of a short-term process E_1 and a long-term process E_2. After a stimulus has been presented, E_1 starts off with an amplitude S_1 and decays with a rate constant $c_1 = \ln(2)/2$ seconds, according to its half-time of 2 seconds, whereas E_2 starts off with an amplitude S_2 and decays with a rate constant $c_2 = \ln(2)/140$ seconds.[6]

$$E(t) = E_1(t) + E_2(t) = S_1 \cdot \exp(-c_1 t) + S_2 \cdot \exp(-c_2 t), \quad t \geq 0 \quad (4)$$

If stimuli are presented at successive times $t_1, t_2, \ldots, t_n$, the total excitatory effect $E(t)$ at any time $t \geq t_n$ is

$$E(t) = \sum_{i=1}^{n} E_1(t - t_i) + \sum_{i=1}^{n} E_2(t - t_i)$$

$$= S_1 \cdot \sum_{i=1}^{n} \exp(-c_1(t - t_i)) + S_2 \cdot \sum_{i=1}^{n} \exp(-c_2(t - t_i)) \quad (5)$$

This process was programmed on an analog computer and, as in the experiment described in Fig. 8, eighteen "stimuli" each triggering a process as written in Eq. (4) were given at 10-second intervals. The result is plotted in Fig. 9 which shows that while the short-term process E_1, represented by

[6] For reasons of simplicity it is assumed that the process starts with its peak at stimulation time $t = 0$. As shown in Fig. 7a, however, this peak is reached somewhat later. This fact is accounted for in the model given in Fig. 13.

FIG. 9. Analog computer process responding to 18 stimuli presented at 10-second intervals. E_1 represents the short-term process, E_2 the long-term process as formulated in Eq. (5). One volt corresponds to a "chirp rate" of 1 per second.

the difference on the ordinate between the upper and lower curve, builds up very little, the long-term process builds up stepwise and leads to a gradual rise in chirp rate persisting for many seconds after stimulation has ceased. The values of S_1 and S_2, required to match the results in Fig. 8, were 0.8 and 0.006.

According to the model outlined above, the long-term buildup in chirp rate should be higher if stimuli are presented at shorter intervals. The data given in Fig. 10 were obtained by presenting stimulus chirps over a 3-minute period at 10-, 5-, 2.5-, 1.25-, or 0.625-second intervals, referred to in the figure as stimulation 1, 2, 3, 4, and 5, respectively.

Since the short-term increment in chirp rate, due to its half-time of 2 seconds, decays within 10 seconds to $\frac{1}{32}$ of its initial value, the value of

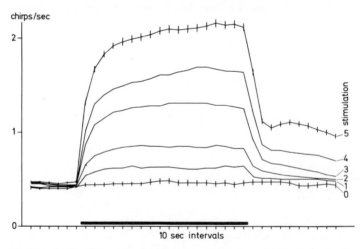

FIG. 10. The rise in chirp rate obtained by a 3-minute period of continual stimulation (bar above abscissa) at five different stimulus rates. The long-term increment seen in the poststimulatory period is larger the higher the stimulus rate, which is 1, 2, 4, 8, and 16 chirps per 10 seconds, referred to as stimulation 1, 2, 3, 4, and 5 respectively. Stimulation 0 refers to the control experiment with no stimuli presented. The curves represent average values of approximately 500 experiments for each type of stimulation at prestimulatory chirp rate levels between 0 and 1 chirp per second. Variances of the mean are given for two curves. (From Heiligenberg, 1969.)

the long-term buildup obtained during a 3-minute period of stimulation can be estimated from the chirp rate recorded 10 seconds after the last stimulus chirp was presented. When the chirp rate, in that poststimulatory 1-minute interval starting 10 seconds after the last stimulus presentation, is plotted as a function of the chirp rate during the 1-minute interval preceding the first stimulus, it is found that the value of the long-term increment depends only on the rate of stimulation but not on the prestimulatory chirp rate level (Fig. 11). This shows that the long-term increment, as was already assumed in connection with Fig. 9, is additive in the same manner as the short-term increment (cf. Fig. 7).

The time constant of the long-term process just described is of the same order as the time constant known for the excitatory processes in attack behavior of cichlid fish as outlined in Sections IV,A and B (Figs. 3 and 4). These processes however, appear to be of a short-term nature compared to a much slower process in attack behavior found recently. When a fish dummy raising the attack rate in a male *Haplochromis burtoni* as described in Section IV,B, is presented many times every day, the attack rate of the male stays at a moderate level. When neither dummies nor adult conspecifics are presented, the attack rate declines to a very low value. This phenomenon led to the assumption that a single dummy presentation, apart from raising

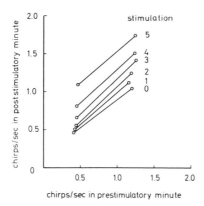

FIG. 11. Average level of chirp rate (ordinate) observed after a 3-minute period of continual stimulation as a function of prestimulatory chirp rate (abscissa) and rate of stimulation (1, 2, 3, 4, 5). The long-term increment in chirp rate, represented by the vertical distance of a particular line from the lowest line depends on the rate of the corresponding stimulation (explained in Fig. 10), but not, within limits, on the prestimulatory chirp rate of the cricket. Data were grouped according to different ranges on the abscissa and average values calculated with respect to both coordinates. Prestimulatory chirp rates beyond 2 chirps per second were not frequent enough to yield reliable average values and therefore were discarded. The poststimulatory minute referred to on the ordinate starts 10 seconds after the last stimulus chirp presentation.

the attack rate of the male for a few minutes, also causes a small but long-lasting increment in attack rate. To study this long-term process quantitatively, the experiment described below was performed (Heiligenberg and Kramer, 1972).

An adult male was placed in an aquarium together with ten small fish, serving as a test stimulus to measure the attack readiness in the male. After having seen neither dummies nor adult conspecifics for several weeks, the attack rate of the male reached a very low level. Then, on 10 successive days, from 900 hours until 1700 hours each day, a dummy resembling a male with a vertical black eye-bar was presented every 15 minutes for 30 seconds. As shown in Fig. 12, the attack rate, recorded for 1 to 3 hours each forenoon, rose during these 10 days of stimulation and returned to its low prestimulatory level with an estimated half-time of 7 days.

The continuous line in Fig. 12 represents a theoretical long-term process, which increases by 0.0015 attacks per minute with each dummy presentation and decays with a half-time of 7 days. These values gave the best least squares fit to the data given in Fig. 12.

These examples demonstrate that a certain stimulus pattern may elicit complicated excitatory processes which can be written as sums of exponential functions [cf. Eq. (4)]. It should be noted that the distinction between "short-term" and "long-term" processes described so far is of only a relative nature and is not based on the absolute values of the time constants involved.

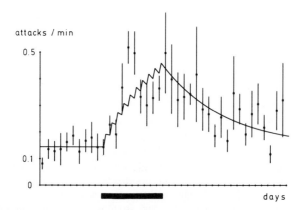

Fig. 12. Long-term effect on attack rate in males of *Haplochromis burtoni*. The average attack rate rises within 10 days of continual dummy presentations (bar underlining abscissa) and returns to its low prestimulatory level with a half-time of 7 days. Data of four individuals with a total of six experiments were averaged, variances of the mean are indicated by vertical bars. Individuals were observed for 1 to 3 hours each forenoon. The continuous line represents a theoretical long-term excitation. (From Heiligenberg and Kramer, 1972.)

The fact that a certain stimulus pattern may initiate a short- and a long-term process of excitation leads to a difficulty in determining the constant additive increment for the short-term process, as shown in Figs. 3b and 7b, for the following reason. Unfortunately, an animal does not always respond to a particular stimulus pattern equally strongly: periods of high responsiveness alternate with periods of very low responsiveness without any obvious external correlates. Fluctuations in responsiveness to chirp stimuli were found not to be correlated with fluctuations in "spontaneous" chirp activity (Heiligenberg, 1966). If an animal happens to be in a period of high responsiveness, presentations of a certain stimulus pattern at intervals not long enough to allow for a sufficient decay of the long-term process, will lead to a gradual rise of the corresponding behavior rate. This in turn results in subsequent infrequent low prestimulatory levels of activity. If however, the animal is in a period of low responsiveness, successive stimulations will hardly raise its activity and many instances of low prestimulatory activity levels will subsequently be found. When data are averaged as in Figs. 3b and 7b, data on the left-hand end of the abscissa are obtained primarily from periods of low responsiveness and consequently the increment found is lower than at higher ranges of the abscissa.

To explain a steady buildup of excitation under repeated stimulation on the basis of Eq. (5), it was assumed that each stimulus causes constant short- and long-term increments S_1 and S_2, the values of which do not decrease under repeated stimulation. This, however, requires that the animal either does not habituate to the stimulus presented repeatedly or recovers between successive presentations. When the short-term increments caused by successive stimuli in Fig. 8 are compared, no significant decrement is found. Therefore one may conclude either (a) that the cricket habituates to successive stimulus chirps so slowly that no significant decrement in response strength can be detected after only 3 minutes of stimulation, or (b) that the process of recovery is so fast that the cricket recovers each time within the 10-second interval separating successive stimuli. In either case, as long as the time constants of habituation (cf. Hinde, 1970) are by order of magnitude different from the duration of the stimulus sequence and the length of the interstimulus intervals, S_1 and S_2 in Eq. (5) can be considered constant. If the time constants of habituation do not fulfill this condition, S_1 and S_2 become functions of the stimulation history in time.

In order to demonstrate a short-term habituation, recovering within 10 seconds as assumed above, the total excitation obtained when stimuli are presented at intervals shorter than 10 seconds must be compared with the corresponding value predicted by Eq. (5). It emerges that as the interstimulus interval becomes smaller, the theoretically predicted increment in chirp rate is larger than actually found in the experiment (Heiligenberg, 1969), even if one takes into consideration the fact that at a higher stimulus rate more chirps are likely to coincide with chirps of the cricket and thus become ineffective. This result permits the conclusion that interstimulus intervals shorter than 10 seconds do not provide sufficient time for recovery to occur. This explanation is backed by the behavior of single auditory units in the first thoracic ganglion, which fire at a lower rate if stimulus chirps are presented at intervals shorter than 10 seconds, whereas almost no habituation is encountered if intervals of at least 10 seconds separate successive stimuli (Heiligenberg, unpublished experiments).

As far as the experiments on cichlids are concerned, 15-minute intervals seem to be sufficiently long for the animal to recover. Even after 10 days of continual stimulation no systematic decrement was encountered.

V.   THEORETICAL CONSIDERATIONS

As shown in this paper, the readiness of a cichlid to attack other fish, measured by the rate of attacks directed at a given number of small test fish, is additively increased if a certain stimulus pattern is presented (Fig. 3b). The same is found for the readiness to chirp in a cricket (Fig. 7b), measured by the rate of chirping. The fact that, within certain limits, a stimulus presentation leads to a constant increment E added to the state of readiness R, does not yet prove that what actually occurs inside the animal is an additive process. If, for example, the readiness to carry out a particular behavior were measured by the exponential function of the rate of occurrences observed rather than by the rate itself, the additive relation among rates would turn into a multiplicative one between exponentials: if R_0 denotes the level of readiness to be expected without stimulation, and R_s the level of readiness obtained after applying a stimulus raising the readiness by the amount E, the additive relation $R_s = R_0 + E$ is equivalent to the multiplicative relation $\exp (R_s) = \exp (R_0) \cdot \exp (E)$, which says that the readiness, measured by $\exp (R)$, is multiplied by the factor $\exp (E)$ when the stimulus is presented.

The mathematical relation between the given state of readiness in the animal and the effect of a certain stimulus presentation thus depends on the type of measure applied, and therefore tells us little about mathematical operations occurring on the physiological level.

Though the actual mathematical relation between R and E depends on the measure applied, the fact that the effect E of a stimulus presentation can be written as a term not depending on the present state of readiness R deserves special attention: it shows that the prestimulatory state of readiness in the animal and the effect caused by a certain stimulus can be considered as two mutually independent processes, which are combined into a new poststimulatory state of readiness. This combination is found to be additive, if the state of readiness is measured by the rate of events.

The finding that the effects of different stimulus patterns, such as the black eye-bar and the field of orange spots are combined additively, represents a quantitative example of the rule of "heterogeneous summation" stated by Seitz (1942) and further investigated by Baerends (1962). This additive relation, however, depends on the measure applied: any nonlinear transformation of this measure will yield a different mathematical relation, as shown above for the relation between E and R.

The current state of readiness R can thus be considered to be a sum of all stimulus effects E_1, E_2, . . . , caused by the appearance of different stimulus patterns and characterized by different initial amplitudes and time constants of decay, along with some intrinsic base-line residual I, which persists if no stimuli have been presented for a sufficiently long period of time (Fig. 13). This residual I determines R under "steady" conditions as studied in Section III, and the serial correlation functions given in Fig. 1 characterize the time course of fluctuations in I.

The theoretical concept outlined in Fig. 13 is supported by other investigations. Thus the readiness of an animal to carry out a particular behavior, measured in this paper by the rate of occurrences of this behavior in a standardized situation, can in some fish species be estimated from the momentary coloration of the animal. This was demonstrated most convincingly by Baerends and his co-workers (1955), who showed that the rate of occurrence of certain items of "courting" behavior in *Lebistes reticulatus* is correlated with the present coloration of the animal, and that different colorations could be ranked according to increasing rates of behavioral performances associated with them. At a given state of readiness of a male to court, referred to as "internal stimulation" and determined by the type of coloration, a large female was more likely to elicit a particular courting pattern than a small female; accordingly, the size of the female presented was taken as a measure of "external stimulation." By determining the minimum external stimulation necessary to elicit a particular courting pattern at different states of "internal stimulation," the authors found that the total amount of internal and external stimulation has to reach a critical value. According to the data given (Baerends *et al.*, 1955, Fig. 24, p. 307), this total amount could be interpreted as the product of external and internal stimulation rather than their sum. However, since the measures of external and internal stimulation are arbitrary, they can in principle be replaced by their logarithms. When this transformation is applied, the multiplicative relation between the original measures of external and internal stimulation turns into an additive relation between the new measures obtained. This result is now in agreement

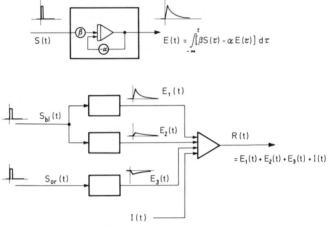

Fig. 13. Analog computer flow diagram representing the state of readiness R as a function of stimulatory inputs S and an intrinsic variable I. The state of readiness, R, measured by the rate of behavioral events in a standardized situation is the sum of an intrinsic variable I and certain aftereffects E caused by the appearance of particular stimuli. The upper part of the diagram shows an analog computer circuit, which transforms the stimulus input S, represented by a short impulse, into an excitatory process E. E decays with the rate constant α after it has reached its peak, which is proportional to β/α. When stimuli are presented in succession, $E(t)$ shows a time course as presented by E_1 and E_2 in Fig. 9, the degree of buildup depending on the rate constant α and the size of the interstimulus interval. The lower part of the diagram represents the attack readiness of a male cichlid which is raised by a short- and a long-term process, E_1 and E_2 respectively, whenever the "vertical black eye-bar" pattern, S_{bl}, is presented, and lowered by another process E_3, whenever the "field of orange spots above pectorals" pattern, S_{or}, is presented.

with the flow diagram shown in Fig. 13, which says that in order to obtain a certain minimum readiness R', the sum of the given prestimulatory level of R and the total stimulatory input $E_1 + \cdots + E_N$ initiated by the stimulus presentation should reach the critical value R'.

The model presented in Fig. 13 considers external stimuli as inputs comparable to short impulses, since all experiments in this paper were aimed at presenting stimuli for relatively short periods of time. Within a natural situation, however, certain stimuli are present for longer periods of time and one may thus ask whether the model in Fig. 13 can still be applied in this case. If one considers that a stimulus is most effective when it appears anew, and that an animal habituates to a stimulus which is constantly present, it appears reasonable to represent the effect of any stimulus as a short impulse marking the moment of its appearance and to neglect its subsequent presence. The effect on a certain state of readiness caused by a stimulus which is generally present for longer periods of time, such as the sex partner of a ring dove (cf. Lehrman, 1965) or a nest cup of a canary (cf. Hinde, 1965) may thus arise by the same mechanism as outlined in Fig. 13.

The values of the time constants characterizing excitatory and inhibitory processes described in this paper raise speculations about their possible physiological origin. Time constants on the order of several minutes have also been found in behavioral and neuronal processes resulting from the presentation of certain visual stimuli or electrical brain stimulation in birds (cf. Harwood and Vowles, 1967; Delius, 1970, 1973).

In order to arrive at a physiological interpretation of the state of readiness, a further phenomenon should be mentioned. As outlined in Section II, a test stimulus has to be applied in order to determine the readiness for a behavior which occurs in response to a certain class of stimuli. Preliminary observations have shown that the rate of attacks directed at small test fish is, within limits, proportional to the number of small fish available in the aquarium. An adult male is most likely to attack a small test fish when it happens to swim close into his range or simply moves from one point of the aquarium to another. It thus appears that certain behavioral features in a test fish are likely to release an attack by the adult male and consequently the rate of attacks elicited should, within limits, be proportional to the number of potential releasers available. The function of the small test fish would thus be to provide attack-releasing stimuli in a random fashion and at a fairly constant rate.

In the light of the phenomena described so far one may speculate about a physiological representation of the state of readiness R. Assuming that the state of readiness is based on a physiological random variable $r(t)$, which may be a fluctuating membrane potential, a hormone concentration, or a spike rate, one could postulate the following:

(a) When $r(t)$ exceeds a particular threshold r_0, the corresponding behavioral pattern is carried out.

(b) In the case of a type of apparently spontaneous behavior, such as chirping, $r(t)$ reaches its threshold r_0 within a period of time Δt with non-zero probability.

(c) In the case of a type of behavior occurring in response to certain stimuli, $r(t)$ is highly unlikely to reach its threshold r_0 within a period of time Δt, unless a relevant stimulus S is provided, raising $r(t)$ by the amount $e(S,t)$. Whenever $r + e$ exceeds the threshold r_0, the corresponding behavior occurs.

(d) A certain stimulus, S_1, may cause a very brief increment, $e(S_1,t)$, and thus trigger a single behavioral response in the manner that small test fish elicit attacks. Another type of stimulus, S_2, may cause a long-lasting increment, $e(S_2,t)$, comparable to an exponential function E in Fig. 13, and thus affect the average level of $r(t)$ for some period of time following stimulation. An increase in $r(t)$ will of course raise the probability of $r(t)$ exceeding a particular threshold and thus make the corresponding behavior more likely to occur. Though $e(t)$ should be expected to be a random process it will be treated in the following simply as a function of time, which represents the average course of the stimulatory effect.

(e) Given a behavioral event as in (c), of duration Δt, and a probability p of $r(t)$ exceeding r_0 within a given period of time Δt, then only one event can occur within this period Δt, and the probability of this occurrence is p. Consequently, the expected number of behavioral events within a long period of time $T = n \cdot \Delta t$ is $n \cdot p$.

(f) Given a behavioral event as in (d), of duration Δt, a brief stimulatory increment e_0 caused by a test stimulus with probability p_0 within a given period Δt, and a probability p of $r(t)$ exceeding $r_0 - e_0$ within this period Δt, then only one event can occur within this period Δt and the probability of this occurrence is $p \cdot p_0$. Consequently, the expected number of behavioral events within a long period of time $T = n \cdot \Delta t$ is $n \cdot p \cdot p_0$. In a standardized situation the expected rate of test stimulus presentations, $n \cdot p_0$, is constant and therefore the expected number of behavioral events is proportional to the probability p.

The expected rate of behavioral events, which yields a measure of the expected level of readiness R, thus is proportional to the probability p of $r(t)$ reaching a particular threshold within a period of time Δt. In the case of a spontaneously occurring type of behavior (e), this threshold is r_0; in the case of a behavior triggered at a constant rate by a test stimulus (f) causing stimulatory increments of size e_0, this threshold is $r_0 - e_0$. As stated in (d), a stimulatory increment $e(t)$ may raise the average level of $r(t)$ for a considerable period of time. Under the assumption that the value of $e(t)$ does not depend on the prestimulatory average level of $r(t)$, e raises r in the same additive manner as the variable E raises R in Fig. 13. Since R is estimated by the rate of behavioral events in a standardized situation and since this rate is changed under stimulation by an amount which is independent of its prestimulatory level, one must assume that the probability p, which is proportional to the rate of behavioral events, is also changed under stimulation by an amount independent of the prestimulatory level of p. This however requires that this probability p is—within limits—a linear function of the average level of $r(t)$.

Some ethological phenomena can be described and their classical interpretation questioned on the basis of the above stochastic formulation. A behavioral pattern ordinarily occurring in response to a certain class of stimuli may occur spontaneously when such stimuli have not been presented for a long period of time. This so called "vacuum activity" (Lorenz, 1937) was often referred to as a result of a so-called "drive" which builds up spontaneously and is reduced by the performance of the corresponding behavior. If this behavior is not released for a sufficiently long period of time this "drive" is assumed to build up to such a high level that the behavior will eventually "go off" without any external releaser. Considering the random process $r(t)$ one would say that although $r(t)$ is extremely unlikely to reach its critical threshold r_0 within a period of time Δt, unless it is raised by a stimulatory increment e, the probability that $r(t)$ will reach its critical threshold "spontaneously" within a long period of time T rises with the value of T. In order to claim any buildup of readiness for this particular behavior under stimulus deprivation, one has to show that the probability of "vacuum" occurrences increases during successive periods of time T_1, $T_2, \ldots$, all $T_i = T$.

Another phenomenon, related to "vacuum activity," is "response threshold lowering" under stimulus deprivation: this refers to cases in which the animal appears to respond to inappropriate stimuli if appropriate stimuli are not presented for a sufficiently long period of time. In terms of a "drive" theory this phenomenon is again explained by a spontaneous buildup of a drive so that less and less effective stimuli become sufficient to release responses, the extreme case being that the behavior "goes off" without any stimulus at all. Considering the random process $r(t)$ one would assume that appropriate stimuli cause a considerably larger increment e than do inappropriate stimuli and that, when both types of stimuli are available, the "strong" stimulus will raise $r(t)$ above its threshold r_0 long before the "weak" stimulus is given any chance. However, when no appropriate stimulus is available during a period of time T, an inappropriate stimulus will raise $r(t)$ above its critical threshold r_0 with a probability that increases with the value of T. In order to claim any buildup of readiness under the absence of appropriate stimuli, one has to show that within successive periods of deprivation T_1, $T_2, \ldots$, all $T_i = T$, a certain inappropriate stimulus is in fact more and more likely to trigger responses.

VI. Ecological Significance

The fact that certain stimulus patterns affect particular behavioral states of readiness raises the question of the advantage such mechanisms have for the animal. This problem should be approached by studying the whole natural behavior of the animal in the context of its ecology. Unfortunately little is known about the behavior of cichlid fish under natural conditions. But due to the fact that these animals thrive under aquarium conditions one should assume that their behavior in captivity reflects many relevant

traits of the behavior in a natural environment, especially if these animals
are kept in very large aquaria providing space for many territories. Considering the aggressive behavior in *Haplochromis burtoni*, the following may be
said about the biological relevance of the mechanisms mentioned.

Juvenile fish live in schools and are uniformly gray. At the age of a few
months the largest fish in the school develop a black vertical eye-bar for
longer and longer periods of time, tend to stay close to the ground in open
areas, and eventually begin to chase other fish away from their areas. Several
males may establish neighboring territories and mutually fight each other
on the boundaries. Within a few days a male begins to dig a pit in the
center of his territory. At this time he develops a field of orange spots above
the pectoral fins.

Females do not develop any color patterns as they grow mature and continue to live in schools. As they get ready to spawn they stay close to a
male's territory and rarely flee when attacked. They finally spawn with a
male and then leave the area carrying the fertilized eggs in their mouths.
After about 10 to 14 days the young fish swarm out of the mother's mouth.
By then the mother has established a small territory in a hidden area, shows
a black vertical eye-bar, and attacks intruders viciously.

On the basis of these observations it appears probable that another fish,
showing a black vertical eye-bar and thus representing a potential rival trying
to find a territory, raises the readiness to attack in a resident male, which
will consequently spend more effort in chasing away other males rather than
juvenile fish or females which do not represent any threat to his territory.
Whereas the short-term increment in attack readiness prepares the owner
of the territory immediately for a single imminent encounter, repeated encounters lead to a considerable long-term buildup of attack readiness which
puts the animal in a favorable position to withstand such encounters as they
become more frequent.

One could now raise the question why the readiness to attack in a territorial male could not spontaneously stay at a very high level without requiring replenishment by external stimuli. For unknown reasons, different states
of readiness may to some extent be mutually exclusive, a high level of readiness for one particular behavior excluding a comparably high level of readiness for another behavior, even if the two behavioral patterns concerned
are not mutually exclusive. It therefore appears more economical to raise
a particular state of readiness whenever there is a need for the corresponding
behavior, rather than keeping this state of readiness at a high level at the
expense of other states of readiness.

The inhibitory effect on the readiness to attack caused by the field of
orange spots above the pectoral fins may serve, as proposed by Leong
(1969), to keep the readiness to attack within a colony of territorial males

at a moderate level. Since this color pattern is only shown by males which have already established a territory, the consequence of this mechanism would be that a neighboring territorial male raises the readiness to attack less than does a young male which is still trying to find a territory and thus represents a more serious threat.

No satisfying explanation can be given so far for a cricket's response to chirp sounds, since little is known about the behavior of these insects in the context of their natural environment. Unless the behavior of an animal has been studied under more natural situations, there is little hope of understanding the biological relevance of behavioral mechanisms investigated under laboratory conditions.

Acknowledgment

I wish to thank John Thorson, Russell Fernald, and Juan Delius for their critical comments on my manuscript.

References

Baerends, G. P. 1962. La reconnaissance de l'oeuf par le Goéland argenté. *Bull. Soc. Sci. Bretagne* **37**, 193–208.

Baerends, G. P., and Baerends-von Roon, J. M. 1950. An introduction to the study of the ethology of cichlid fishes. *Behaviour, Suppl.* **1**.

Baerends, G. P., Brouwer, R., and Waterbolk, H. T. 1955. Ethological studies on *Lebistes reticulatus*. I. An analysis of the male courtship pattern. *Behaviour* **8**(4), 249–334.

Delius, J. D. 1970. The effect of daytime, tides and other factors on some activities of lesser blackbacked gulls, *Larus fuscus*. *Rev. Comp. Anim.* **4**, 3–11.

Delius, J. D. 1973. Agonistic behaviour of juvenile gulls, a neuro-ethological study. *Animal Behav.* **21**(2), 236–246.

Harwood, D., and Vowles, D. M. 1967. Defensive behaviour and the after effects of brain stimulation in the ring dove (*Streptopelia risoria*). *Neurophysiologia* **5**, 345–366.

Heiligenberg, W. 1965. The effect of external stimuli on the attack readiness of a cichlid fish. *Z. Vergl. Physiol.* **49**, 459–464.

Heiligenberg, W. 1966. The stimulation of territorial singing in house crickets. *Z. Vergl. Physiol.* **53**, 114–129.

Heiligenberg, W. 1969. The effect of stimulus chirps on a cricket's chirping. *Z. Vergl. Physiol.* **65**, 70–97.

Heiligenberg, W., and Kramer, U. 1972. Aggressiveness as a function of external stimulation. *J. Comp. Physiol.* **77**, 332–340.

Heiligenberg, W., Kramer, U., and Schulz, V. 1972. The angular orientation of the black eye-bar in *Haplochromis burtoni* (*Cichlidae, Pisces*) and its relevance to aggressivity. *Z. Vergl. Physiol.* **76**, 168–176.

Hinde, R. A. 1965. Interaction of internal and external factors in integration of canary reproduction. *In* "Sex and Behavior" (F. Beach, ed.). Wiley, New York.

Hinde, R. A. 1970. Behavioural habituation. *In* "Short-term Changes in Neural Activity and Behaviour" (G. Horn and R. A. Hinde, eds.), pp. 3–40. Cambridge Univ. Press, London and New York.

Lehrman, D. S. 1965. Interaction between internal and external environments in the regulation of the reproductive cycle of the ring dove. *In* "Sex and Behavior" (F. Beach, ed.). Wiley, New York.

Leong, C. Y. 1969. The quantitative effect of releasers on the attack readiness of the fish *Haplochromis burtoni* (*Cichlidae*). *Z. Vergl. Physiol.* **65,** 29–50.

Lorenz, K. 1937. Über die Bildung des Instinktbegriffes. *Naturwissenschaften* **25,** 289–300, 307–318, 324–331.

Seitz, A. 1940. Die Paarbildung bei einigen Cichliden. I. *Z. Tierpsychol.* **4,** 40–84.

Seitz, A. 1942. Die Paarbildung bei einigen Cichliden II. *Z. Tierpsychol.* **5,** 74–100.

Time-Sharing as a Behavioral Phenomenon

D. J. McFarland

DEPARTMENT OF EXPERIMENTAL PSYCHOLOGY
OXFORD UNIVERSITY
OXFORD, ENGLAND

The complexity of the temporal organization of behavior has inspired many different methods of approach (e.g., Baerends *et al.*, 1955; Cane, 1961; Wiepkema, 1961, Hinde and Stevenson, 1969). The present approach differs from these in that it is classificatory rather than descriptive or explanatory in nature. The transition from one activity to another is taken as the fulcrum of inquiry, and an attempt is made to distinguish different classes of activity from a theoretical standpoint, and to show how these may be identified empirically.

I. Competition and Disinhibition

A motivational system can be defined on the basis of the assumption that the total behavior repertoire of an animal is composed of a number of distinct categories of behavior, each of which is controlled by one of a number of separate systems. Such a definition is essentially arbitrary and chosen for convenience. In view of the complexity of interactions between such systems, it is doubtful that a rigorous separation is possible, but a degree of separateness of factors controlling different categories of behavior must exist (McFarland, 1971), and it is useful to take separability of motivational systems as a working hypothesis. A motivational system can thus be envisaged as a system controlling a group of functionally related activities. It is convenient to refer to such systems in general terms, as "feeding system" or "aggression system."

At any particular time, many motivational systems will be active, and these must in some sense compete for the *behavioral final common path,* so called because it involves the last type of interaction in the causal chain, the final decision before the potential activity becomes overt. This term is strictly relevant to consideration of motivational systems in behavioral terms, rather than to the structure of the nervous system. Von Holst and von Saint Paul (1963) used the term "initial common path" for competition at the perceptual level, and "final common path" (after Sherrington, 1906) for competition at the motor level. Both types of competition are envisaged as operating in the behavioral final common path of a motivational system (McFarland, 1971).

The level of causal factors for some systems will at any time, be greater than that of others, so that the animal may be said to have a set of motivational priorities. Thus feeding might be the top priority activity, grooming the second in priority, and sleep the third. We can expect the order of priority to be related to the sequence of overt activities that are observed in a particular situation, especially when there is little change in external stimulation. The question is: what factors determine which motivational system is to have priority in the behavioral final common path, and thus gain overt expression?

The simplest answer to this question is that the system which has the highest level of causal factors gains priority by virtue of competition with the other systems. The other systems are then said to be subject to behavioral inhibition. According to Hinde (1970), "Behavioural inhibition is . . said to occur when the causal factors otherwise adequate for the elicitation of two (or more) types of behaviour are present, and one of them is reduced in strength because of the presence of the causal factors for the other. . . . In practice, since it is usual for causal factors for more than one type of behaviour to be present, some degree of behavioural inhibition probably occurs all the time." The amount of time an animal spends at a particular activity may thus be restricted by the necessity for doing other things. For instance, the proportion of the day spent feeding by blue tits increases from about 70% in summer to 90% in winter, and there is a correlated decrease in the time spent resting and preening (Gibb, 1954). Cotton (1953) studied the effect of food deprivation on running speed in rats trained to run for food reward. Overall running speed was found to increase with increased deprivation time, but the effect was much less marked when the time spent at "competing responses," such as grooming and sniffing, was subtracted from the overall score. Thus the most important effect of deprivation was the increased priority of running for food. A similar effect was found by Cicala (1961).

Taken at face value, these findings suggest that the causal factors for a particular activity build up to a level sufficient to oust the ongoing behavior by motivational competition. Continued performance of a behavior pattern produces consequences which feed back and, if their effect is negative, the overall level of causal factors may be reduced to a point where the causal factors relevant to another activity are sufficiently strong to take over. Thus each motivational system follows a kind of relaxation oscillation, each "drive" finding its own level in a "free market" (Logan, 1964). However, far from being self-regulatory, some motivational systems are partly under the control of permissive factors from other systems. Ad libitum drinking, for instance, is dictated primarily by food intake in rats (Fitzsimmons and Le Magnen, 1969; Kissileff, 1969) and doves (McFarland, 1969).

Animals rarely indulge in one type of activity exclusively for long periods of time. Feeding, courting, running in a maze, are generally interspersed with other activities which interrupt the primary ongoing behavior. Many psychologists call such activities "competing responses" on the assumption that they arise by motivational competition, as outlined above. However, *a change in behavior due to competition can in practice be recognized when a change in the level of causal factors for a second-in-priority activity results in an alteration in the temporal position of the occurrence of that activity* (McFarland, 1969). In other words, a change from activity A to activity B should occur earlier when the level of causal factors for B is high (but not higher than that of A), than when it is low. Unless such a test is applied, the term "competing response" has descriptive validity only. Falk (1969) prefers the term "adjunctive behavior" since it implies only that the observed behavior occurs as an adjunct to a particular situation without evaluation of its causation.

Competition is characterized by the level of causal factors for a particular activity being ultimately responsible for the removal of behavioral inhibition on that same activity. The alternative is that the inhibition is removed by other factors, so that the causal factors for a particular activity play no role in the removal of inhibition on that activity. This is the essence of disinhibition. Behavioral disinhibition as a mechanism responsible for the occurrence of adjective behavior was originally (e.g., van Iersel and Bol, 1958; Rowell, 1961; Sevenster, 1961) envisaged as a consequence of conflict between incompatible activities. At the equilibrium point in a conflict the inhibition which the conflicting activities would normally exert on other behavior patterns is removed, allowing a third (displacement) activity to "show through" (Rowell, 1961). There is evidence that such displacement activities occur when the competing predominant behaviors are equally balanced (Rowell, 1961; Sevenster, 1961), thus identifying the equilibrium point as the permissive factor in the situation. However, it is evident that conflict is not necessary for disinhibition to occur. Disinhibited activities can be identified in thwarting (McFarland, 1966a), feeding (McFarland and L'Angellier, 1966), and drinking (McFarland, 1970) situations, and probably occur in many situations where the behavior sequence is stereotyped (McFarland, 1969).

As a general rule, *the time of occurrence of a disinhibited activity is independent of the level of causal factors relevant to that activity* (McFarland, 1969). A simple experiment may serve to illustrate this point. Barbary doves are food deprived and allowed to work for food in a Skinner box. After 5 to 10 minutes the feeding is interrupted by a bout of grooming behavior, and then feeding is resumed. To test whether the grooming is due to competition or to disinhibition, the experiment is repeated, under the same conditions, except that each bird is placed in the Skinner box with a paper clip fastened to the primary feathers on each wing. The results are the same as on the previous occasion, except that the amount and vigor of preening is enhanced. In other words, raising the level of causal factors for preening, by means of paper clips, does change the intensity of the preening behavior, but does not alter its time of occurrence. In this situation preening is a disinhibited activity (McFarland, 1970).

A number of workers have shown that adjunctive behavior can be influenced by manipulation of the relevant causal factors. Thus Rowell (1961) showed that the intensity of displacement grooming in chaffinch could be increased by treating the plumage with water, or dirtying the bill with a sticky substance. Similarly, Sevenster (1961) showed that displacement fanning in the three-spined stickleback could be facilitated by increasing the carbon dioxide concentration in the region of the nest, and McFarland (1965) found that displacement feeding in doves was affected by the presence of grain and the degree of hunger. However, this type of evidence alone is not sufficient to show that disinhibition has occurred. It is also necessary to show that the temporal occurrence of the activity in question is not affected by such treatment. This type of evidence was used by Sevenster (1961) in concluding that courtship fanning in the stickleback is a disinhibited activity, and by McFarland (1970), who showed that adjunctive behavior in feeding and drinking situations is primarily disinhibited.

The mechanisms responsible for behavioral disinhibition are by no means understood, and the disinhibition phenomenon poses considerable problems for theories of behavior control. In particular the fact that the frequency of occurrence of behavior relevant to one system can come under the control of other motivational systems, means that motivational interactions at the level of the behavioral final common path are much more complex than might appear at first sight. However, the present concern is less with the causation of disinhibited behavior, than with its identification and functional relevance.

II. Behavior Sequences

The demonstration that a particular change in behavior is due to disinhibition does not necessarily mean that the disinhibited activity remains under the control of the disinhibiting system. When a dominant system permits the occurrence of an activity relevent to another system, it may thereby lose control and cease to be dominant. Alternatively, it may retain control and reestablish itself after a particular period of time.

This distinction can be made more clearly by reference to Fig. 1. In this figure, three possible explanations of a simple alternating behavior sequence are described in terms of two basic types of behavioral transition. An activity may either be terminated by inhibition from causal factors relevant to another activity, or it may be self-terminating and thus disinhibit another activity (Fig. 1). A simple competition model implies that every transition is of the former type (Fig. 1). Miller's (1944) analysis of approach-avoidance

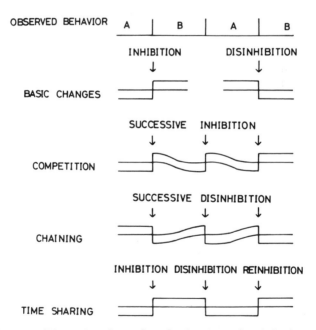

Fig. 1. Three possible explanations of a simple alternating behavior sequence, in terms of two basic types of behavioral transition. Lines represent schematized levels of causal factors as a function of time.

conflict provides an example of this. Approach and avoidance tendencies are seen as competing "drives," the overt behavior being determined by the stronger drive.

An alternative to the competition paradigm is the type of sequence in which the behavioral transitions occur as a result of successive disinhibition (Fig. 1). Such disinhibition may occur as a consequence of the behavior, either through the action of internal stimuli as in satiation of feeding or through external "consummatory stimuli." An example of the latter is provided by the phases of courtship of the three-spined stickleback (*Gasterosteus aculeatus*). As originally envisaged by Tinbergen (1951), these depend upon the succession of stimuli which each partner presents to the other (Fig. 2). Each new set of stimuli may raise the level of causal factors for a particular type of behavior and inhibit the previous behavior. Thus the swollen belly and posture of the female may elicit the zigzag dance of the male. The relevant point is that the time of occurrence of a behavioral transition is dependent upon factors controlling the ongoing behavior in the case of disinhibition, but is independent of the ongoing behavior in the case of competition. It may be argued that this distinction is blurred in cases where stimuli indirectly dependent upon the ongoing behavior play a role in determining the transition. For example, in stickleback courtship the stimuli provided by the female may be partly consequent upon the behavior of the male. The counterargument hinges upon the use of the word "control." A behavioral transition may be said to be controlled by a particular system when the transition time is a function of the behavior of that system. Thus the definitions of competition and disinhibition (see above) are directly related to the empirical test. If manipulating second-in-priority causal factors has no effect upon the transition time, then the transition is due to disinhibition, by definition. In the stickleback case, each activity is supposed to be dependent upon consequences of the previous activity, so that each transition would be classified as disinhibition.

A third possibility is that an activity established by inhibition is self-terminating, thus disinhibiting another activity, but reappears by reinhibition after a particular period of time (Fig. 1). In this case, not only does the dominant system determine the time of occurrence of the disinhibited activity, but it also determines its duration. This phenomenon may be called time-sharing, because the dominant system permits alternative behavior for a certain time period, thus sharing the behavioral final common path with other motivational systems.

Rowell (1961) suggested that one of the factors controlling displacement preening, observed during approach-avoidance conflict in chaffinches, was the time for which equilibrium between approach and avoidance was maintained. Rowell observed that displacement grooming was frequently inter-

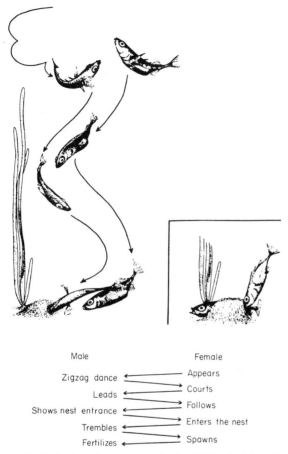

Male   Female

Zigzag dance ←———————— Appears
Leads ←———————— Courts
Shows nest entrance ←———————— Follows
Trembles ←———————— Enters the nest
Fertilizes ←———————— Spawns

Fɪɢ. 2. The mating behavior of the three-spined stickleback (above), and schematic representation of the relations between the male and female (below). (From Tinbergen, 1951.)

rupted by locomotion, but that the time taken to perform grooming actions was generally less than the observed difference between grooming and non-grooming pauses. He found that the probability of grooming varied directly with the average pause length, even though no grooming occurred in the majority of pauses. Rowell's notion of "available time" is a clear example of time-sharing, because both time of occurrence and duration of the disinhibited activity are under the control of the dominant system(s). However, the term time-sharing is, perhaps, more appropriate in situations in which time for alternative behavior is repeatedly allowed. In such cases, it is useful to distinguish between dominant and subdominant motivational systems.

Thus subdominant behavior is that which occurs by permission of the dominant system.

A.   EVIDENCE FOR TIME-SHARING

1.   Circumstantial Evidence

Observations of the type reported by Rowell (1961) can provide circumstantial evidence for the existence of time-sharing. That is, the hypothesis provides a plausible explanation of the observations, but is not tested directly. Similar evidence was obtained by McFarland and Lloyd (1973) in a study of feeding and drinking in doves. When both food and water are available in an operant situation, a hungry and thirsty dove will alternate between feeding and drinking. The frequency of alternation, in terms of quantities ingested, depends largely upon the forced reward rate. However, the probability distribution of the intervals between the beginning of each feeding bout remains the same over a wide range of reward rates (Fig. 3). In other words, the overall average time interval from the beginning of one feeding bout to the start of the next remains the same, even when the rate of ingestion wanes, strongly suggesting that time-sharing is involved.

2.   Pattern of Dominant Behavior

When a behavior sequence is organized on a time-sharing basis, the opportunity and motivation for a particular type of subdominant behavior

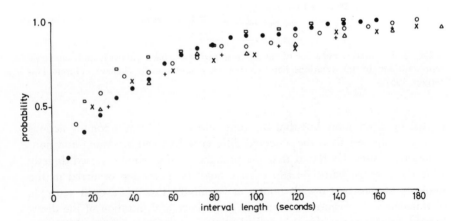

FIG. 3. Cumulative probability distribution of the intervals between the beginning of successive feeding bouts in an operant feeding and drinking situation. Symbols represent reward rates of 15 (●), 10 (○), 7.5 (□), 6 (×), 5 (△), and 4.3 (+) rewards per minute.

should make no difference to the pattern of dominant behavior. Clearly, where a disinhibited activity is self-terminating, its causal factors are likely to influence its duration and thus disrupt the pattern of the disinhibiting behavior. Conversely, when a behavior pattern can be shown not to be affected by manipulation of causal factors relevant to alternative behavior, then the latter is unlikely to be self-terminating. Examples of this phenomenon are reported by McFarland (1970), who found that the feeding satiation curves of doves, in a situation where they characteristically feed and drink alternately, was not affected by the availability of water or by the degree of thirst. In these cases hunger was known to be dominant, but this type of evidence can be used as a test for dominance.

Similar results have been obtained in rats. Six laboratory rats, deprived of both food and water, were tested in a Skinner box with both food and water available, or with only food available. Each animal was used as its own control, and for every subject the feeding satiation curves were similar, whether or not water was available. An example from a single subject is illustrated in Fig. 4. These results suggest that feeding was the dominant behavior in this situation, and that the adjunctive drinking was permitted on a time-sharing basis.

3. Masking Experiments

Some more direct tests for time-sharing are illustrated in Fig. 5. The top diagram (Fig. 5,I) illustrates the basic time-sharing situation. The relative levels of causal factors relevant to two motivational systems are indicated by the lines a and b. Initially, a is ordinate and b is subordinate, and activity A is observed. The level of a then falls, thus disinhibiting b so that activity

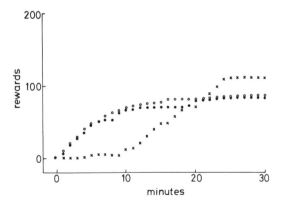

FIG. 4. Cumulative intake in an operant feeding and drinking situation in rats tested with (○) and without (●) water available; (×) indicates water intake.

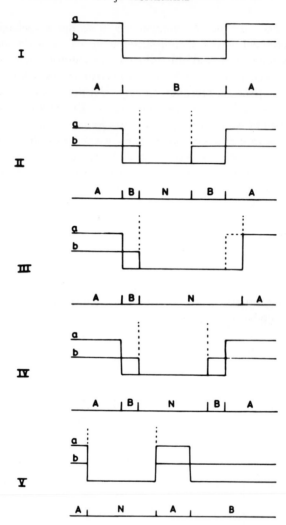

Fig. 5. Possible effects of interruption (shown by dotted lines) in a time-sharing situation, shown as a function of time. Levels of causal factors are indicated by *a* and *b*, while A and B indicate the observed behavior. N indicates that neither A nor B is observed. Note that amount A of B is reduced in IV, but not in V.

B is observed. After a period of time *a* reinhibits *b*, and activity A is observed again. The problem is to show that the duration of activity B is dependent on the system controlling A.

The basis of masking experiments is to set up a situation in which both A and B are prevented from occurring. The manner in which this is done depends upon the experimental situation, but any form of interruption which does not affect the levels of *a* and *b* will suffice. The assumption behind the method is that the temporal organization of the behavior persists when the interruption occurs during subdominant behavior, but not when it occurs during dominant behavior. In other words, because the type of subdominant behavior is independent of the dominant system, interruption of subdominant behavior has no effect on the dominant system, but simply masks the subordinate behavior by using up the available time. Interruption of dominant behavior, however, postpones the control operation, which continues when the interruption is over. Thus interruption of dominant behavior should always be followed by a resumption of that behavior, provided that the interruption itself does not alter the causal factors of the motivational systems involved.

a. *Transition-Dependent Blocking.* The basis of this method is to detect a transition from activity A to activity B (Fig. 5,II) and then to block both A and B by a suitable interruption procedure. If the duration of the interruption (blocking duration) is small compared with the time available for activity B (disinhibition duration), then the animal should return to activity B when the interruption is over (Fig. 5,II). If, on the other hand, the blocking duration is long compared with the disinhibition duration, the animal should return to activity A, the opportunity for B having been lost (Fig. 5,III). This method can be exemplified by a simple experiment.

The subject of this experiment was a single male stickleback (*Gasterosteus aculeatus*), which was maintained in reproductive condition in a tank 93 cm long (Fig. 6). The fish had a nest at one end of the tank and would readily court a female placed in a glass jar at the other end. The tank was divided into three compartments of equal size by means of opaque partitions, each having a doorway (10 × 10 cm). The doorways could be closed by remotely controlled opaque sliding doors. The partitions remained in the

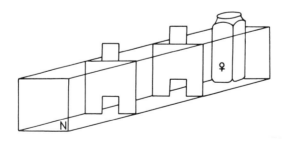

Fig. 6. Testing tank used in stickleback experiment; ♀ indicates female stickleback in a glass jar; N indicates nest of male stickleback. Doorways in opaque partitions can be closed by remote control.

tank for the duration of the experiment, and the male soon habituated to their presence.

When a female was presented, the characteristic behavior of the male was to court the female, attacking through the glass and zigzagging. After a period of courtship the male would swim through both doorways to the nest, where it would often fan the nest, creep through, or manipulate nest material, in the manner described by Sevenster (1961). It would then swim back and resume courtship. This alternating behavior pattern would be kept up for about 15 minutes at a time, after which the male appeared to lose interest. At this point the female would be removed and the experimental session ended. Testing sessions were always separated by a number of hours and were carried out over a period of 5 days.

The first phase of the experiment consisted simply in recording the amount of time spent by the male in the nest (N), middle (M), and female (F) compartments. This phase lasted until 33 nest visits had been recorded. In the second phase a routine procedure was used. When the male entered the M compartment, it was trapped there for a specified period by closing both doors. The trapping periods were 1, 5, 10, 15, 20, and 30 seconds, covering the range of time the male was observed to spend in the N compartment during the first phase of the experiment. Trapping was carried out both when the male entered the M compartment from the N side (N–F transition), and from the F side (F–N transition). There were thus 12 possible combinations of transition-direction and trapping period. The male was tested 10 times at each of these combinations, making 120 trials in all, and these were arranged in random order throughout the experiment. In addition, a "free visit" was allowed each time between trials, during which there was no interference with the behavior of the male. When trapped in the M compartment, the male swam around without obviously attempting to pass through either door, or appearing "agitated" in any way. At the end of each trapping period the doors were opened, and a note made of the compartment (F or N) which the male then entered.

The results of this experiment are illustrated in Fig. 7. The left-hand graphs indicate the times spent in the N and F compartments, recorded in the first phase of the experiment. These are expressed as cumulative frequency distributions (ogives). The percentage of time intervals less than a particular value (abscissa) can be read directly. Thus 50% of the time intervals spent in the N compartment were less than 15 seconds. The right-hand graphs show the percentage of occasions that the fish reversed direction after being trapped in the M compartment. In the case of the F–N transition, the distribution of the percentage of returns to the F compartment is very similar to the cumulative distribution of periods spent in the N compartment, recorded during the first phase of the experiment. Thus the fish is likely to enter the N compartment after being trapped for a short time but

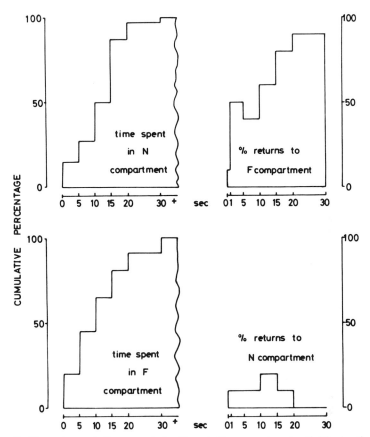

FIG. 7. Percentage of time spent in N and F compartments (left), in the first phase of the experiment, compared with percentage of returns to the F and N compartments (right) in the second phase.

more likely to enter the F compartment following a longer period of trapping. This is precisely the result that would be expected if courtship were the dominant activity, and nest visits were permitted on the time-sharing basis.

When the fish is trapped during an N–F transition, it is likely to enter the F compartment, whatever the trapping duration. This result is consistent with the view that interruption of dominant behavior should be followed by resumption of that behavior.

This experiment provides an example of the type of procedure that can be used to test for time-sharing by the method of transition-dependent blocking. Only one subject was tested in this experiment, but M. t'Hart, working in Leiden, has obtained the same result with a further four sticklebacks and has also obtained similar results in a situation in which the male could see the nest during the period of interruption (personal communication). Clearly, further work must be done before time-sharing can be taken as typical of stickleback courtship, but these experiments do show that time-sharing can occur in this type of situation.

b. Titration. Titration is a development of the method of transition-dependent blocking outlined above. When a transition is detected, behavior is interrupted for a specific period P. If the animal resumes the behavior, the next P is made longer. If it reverts to the former behavior, P is made shorter. Thus the duration of P is titrated. A titration experiment can result in either a stable or an unstable titration. A stable titration is characterized by the fact that the value of P moves up and down around a particular value. During unstable titrations the value of P becomes progressively larger. Two conditions are necessary for a titration to be stable: (1) The titration must be applied to a time-sharing situation, in which certain activities are subdominant. When there is no subdominant behavior, the duration of P will become progressively longer, because interruption of dominant behavior is always followed by resumption of that behavior. (2) The titration must be programmed so that the interruption occurs after a dominant-subdominant transition. Interruption following a subdominant-dominant transition will always result in an unstable titration, because interruption of dominant behavior is always followed by resumption of that behavior. The titration method is exemplified by the following experiment.

Six Barbary doves (*Streptopelia risoria*) were trained to obtain food and water in a Skinner box, in accordance with previously described procedure (McFarland, 1970). They were trained to peck at a red illuminated key to obtain food rewards, and a green key to obtain water rewards. When the keys were unilluminated no rewards were given, and the birds soon learned not to peck unilluminated keys.

The birds were tested following deprivation of both food and water, arranged so that they were either primarily hungry (24-hour food and water deprivation), or primarily thirsty (24-hour water deprivation, 4-hour food deprivation). Under such conditions doves characteristically alternate between feeding and drinking until satiated (Fig. 8). When tested primarily hungry, the behavior was interrupted following all feeding-drinking transitions, and when primarily thirsty they were interrupted after all drinking-feeding transitions. The interruption procedure was to turn off all key illumination and reward mechanisms for the specified period P. Interruptions were instigated at the first peck indicating a transition from dominant to subdominant behavior, and the birds received no reward for that peck. After the specified period P the keys were reilluminated, and a note made of which key the bird then pecked. If this peck indicated dominant behavior, the

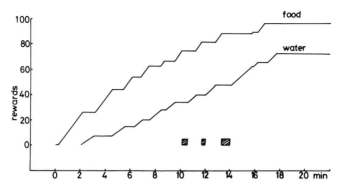

Fig. 8. Cumulative food and water intake in an operant situation. A typical result from a single subject. Shaded blocks indicate nonfeeding and drinking behavior.

duration of P was halved on the next interruption. If the peck indicated subdominant behavior, P was doubled. For example, a primarily hungry animal would be interrupted following a feeding-drinking transition. If it returned to feeding after the interruption, P was halved, etc. This procedure was carried out by a digital computer, which received signals from all pecks and controlled key illumination and reward delivery. A complete record was kept of the complete pecking sequence, and of the durations (P) of all interruptions. Each bird was tested for a 30-minute period under each condition. Titration tests were alternated with noninterrupted sessions, to diminish the possibility of learned modification of the behavior resulting from the interruption procedure.

The results showed that all birds, whether tested primarily hungry or primarily thirsty, gave stable titrations. This is the result that would be expected from previous evidence that time-sharing is involved in this type of situation (McFarland, 1970, 1971; McFarland and Lloyd, 1973).

A convenient way to illustrate the results is on a motivational state plane (McFarland, 1971), in which degree of hunger is plotted against degree of thirst, in terms of grams needed to achieve satiation. An example is illustrated in Fig. 9. As the bird accumulates food and water rewards, the point representing jointly the states of hunger and thirst traces a trajectory as it moves toward the zero point. Each interruption is marked on the trajectory. When an interruption is followed by resumption of the interrupted behavior, the trajectory changes direction. There is no change in direction when interruption is itself instigated at the commencement of each feeding or drinking bout, no reward being given. For example (Fig. 9), a primarily hungry bird which is feeding pecks the water key thus inducing an interruption (marked 1). No water reward is given, and as the bird returned to feeding, the trajectory continues in the same direction. In accordance with

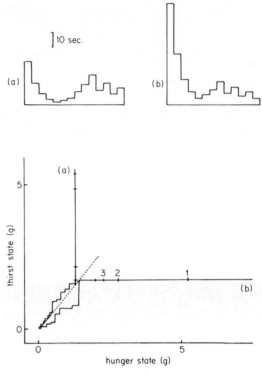

Fig. 9. Titration graphs (above) and trajectories in the hunger–thirst state plane (below). Results from a single subject tested primarily hungry and primarily thirsty. Dotted line indicates boundary of hunger–thirst dominance. Titration graphs show duration of successive interruptions. Trajectories indicate successive motivational states resulting from food and water ingestion. The states are zero when the animal is satiated. The points at which interruptions occurred are marked on the trajectories.

the titration schedule, the duration of the next interruption is halved. A record of the successive interruption periods is shown as a titration graph (Fig. 9).

In Fig. 9 the results obtained from a single bird are shown. The titration graphs and trajectories obtained from both primarily hungry and primarily thirsty conditions, are illustrated in the one figure. Both trajectories home in to the origin, but they do not cross each other and are shown separated by a dotted line. The following argument suggests that this line must be the boundary line between hunger-dominant and thirst-dominant regions of the motivational state plane: (1) There must be a boundary somewhere in the state plane, on one side of which hunger state is greater than thirst state, and vice versa. (2) A titration can only be stable if the trajectory remains on one side of this boundary, since the motivational dominance is reversed with respect to the titration schedule, once the boundary is crossed. (3) It is observed (Fig. 9) that two stable trajectories converge in the state plane, and that a dividing line can be drawn between them. (4) As these two trajectories represent titrations of opposite hunger-thirst dominance, the dominance boundary must be in the region of the dividing line.

The titration method can be used to detect the position of boundaries in the motivational state space, by noting the points at which changes in stability occur. This approach is described by McFarland and Lloyd (1973). In general, time-sharing has the effect of maintaining the trajectory within the dominant region of the state space, but special methods can be used to induce this trajectory to cross a boundary. In addition to providing a tool for motivation analysis, this phenomenon suggests a number of interesting functional considerations, which are discussed later in this paper.

c. Transition-Independent Blocking. In general, interruptions occurring during subdominant behavior mask it, while interruptions of dominant behavior postpone it (Fig. 5, IV and V). Therefore interruptions at random intervals should suppress subdominant behavior, but not dominant behavior. This method was tested in the feeding and drinking situation described above.

Eight doves, which had been used in this type of experiment before, and were familiar with being interrupted in the experimental situation, were used for this experiment. Before testing they were made either primarily hungry or primarily thirsty, in the manner described in the titration experiment. Each bird was allowed to work for food and water for a half-hour period, during which its behavior was interrupted by the method described in the previous experiment. Interruptions were preprogrammed to occur at random intervals, and to be of random duration (range 64 seconds). The total interrupted time was half the total testing time. Control tests were run in which each animal was not interrupted, but could obtain rewards only at a lower rate, calculated to match the overall reward rate in the experimental condition. The reward rate was controlled by introducing a 5-second time-out period after each reward, as described by McFarland (1970).

The results are summarized in Table I. They indicate that when interrupted primarily hungry, drinking behavior was suppressed, and vice versa. This finding supports the view that interruption of dominant behavior postpones it, while interruption of subdominant behavior masks it.

4. Discussion

The evidence discussed above indicates that time-sharing can occur as a behavioral phenomenon. The question of the generality of the phenomenon is an empirical matter, which requires further investigation. Although

TABLE I

FOOD REWARDS EXPRESSED AS A PERCENTAGE OF THE TOTAL REWARDS
OBTAINED IN AN OPERANT FEEDING AND DRINKING SITUATION

Condition when tested	Control	Interrupted	Walsh
Primarily hungry	47.8	81.3	$P < 0.004$
Primarily thirsty	16.4	12.2	$P < 0.027$

the present evidence is primarily based on study of two species, and two behavioral situations, the differences between these situations gives some justification for speculation concerning the generality of time-sharing and its behavioral significance. Some idea of the generality of time-sharing can be gained from the literature on displacement activities. In addition, consideration of the possible significance of time-sharing, in those situations which have so far been studied experimentally, may help to indicate the general type of situation in which time-sharing is likely to be important. The rest of this paper is concerned with these two considerations.

B. DISPLACEMENT ACTIVITIES AS COMPONENTS OF BEHAVIOR SEQUENCES

Historically, the finding that displacement activities can be influenced by manipulation of subordinate causal factors provided the main fuel for disinhibition theories. For example, Rowell (1961) showed that displacement grooming in chaffinches could be increased by treating the plumage with water or by dirtying the bill with a sticky substance. Similarly, Sevenster (1961) showed that displacement fanning in sticklebacks could be facilitated by increasing the carbon dioxide concentration in the region of the nest.

However, this type of evidence alone is not sufficient to show that disinhibition has occurred. It is also necessary to show that the temporal occurrence of an activity is not affected by such treatment. This condition has not always been explicitly met in the displacement literature, though it has sometimes been implicit. However, Sevenster (1961) did note that the frequency of bouts of courtship fanning was not altered by facilitation of fanning with carbon dioxide, except when the tendency to fan was very high. He equated the number of fanning bouts with the frequency of effective disinhibitions and concluded that the frequency of effective disinhibitions is not dependent on the strength of the factors activating fanning.

To account for the increased frequency of fanning observed when these factors were very strong, Sevenster introduced the notion of incomplete disinhibitions which only show up when fanning motivation is high. Personally,

I find this notion unnecessary. A corollary of the disinhibition hypothesis is that a sufficient increase in subordinate factors will result in competition. Indeed, it would be worrying if, in any particular case, the diagnostic criteria for disinhibition did not break down under such conditions. Therefore, I suspect that the observed increase in fanning activity, when fanning is strongly activated, is due to motivational competition and not to disinhibition.

A second point, not always clear in the displacement literature, is whether a displacement activity is dominant or subdominant. Theoretically, it can be either. In the case of Rowell's study of chaffinch grooming it is clear that the displacement grooming, observed during approach-avoidance conflict, must be subdominant if it is true that its duration is limited by the length of the pause.

In the case of Sevenster's study of stickleback fanning, the situation is more complicated. Sevenster concluded that an unfacilitated bout of fanning comes to an end with reinhibition from the sexual system. That is, it is subdominant. Facilitated fanning, however, occurs in longer bouts, and is probably self-terminating. It seems that the time of occurrence, but not the duration, of facilitated fanning is dictated by the dominant system. Therefore such fanning is disinhibited, but not time-shared.

In general, it appears that displacement activities are frequently disinhibited, but may or may not be time-shared. There is good evidence (Rowell, 1961; Sevenster, 1961) that the equilibrium point in a conflict situation can be the permissive factor leading to disinhibition. Displacement activities have also been shown to occur in situations involving physical thwarting and nonreward, and McFarland (1966a,b) suggested that disinhibition was also involved in such situations. Although little direct evidence was provided in these reports, subsequent work (unpublished) suggests that the time of occurrence of adjunctive (displacement) activities in a nonreward (extinction) situation is not affected by causal factors relevant to the adjunctive behavior. For example, extinction curves for feeding in doves are not affected by the presence or absence of drinking cues during extinction, although drinking behavior does occur when the cues are present. Moreover, McFarland (1966a) did find that the course of extinction of drinking behavior was not affected by the presence of grain on the floor of the test cage, although displacement pecking by doves during extinction was affected. Thus in nonreward situations there is some evidence for true disinhibition of displacement activities as judged by temporal criteria. The finding that the temporal pattern of dominant behavior is the same, whether or not displacement activities occur in a particular situation, provides evidence (see Section II,A,2) that time-sharing may also be involved in such nonconflict situations.

In general, evidence for the disinhibition of displacement activities has been obtained for a variety of species and situations, but the evidence for displacement activities involving the time-sharing principle is less good. It seems that the duration of a displacement activity may, or may not, be under the control of the dominant motivational system(s).

C. Functional Consideration

The phenomenon of time-sharing raises a number of questions concerning its functional significance. The advantages of time-sharing over other means of organizing priorities in the behavioral final common path are not obvious. Speculation at this early stage can only be extremely tentative, but perhaps some clue to the functional significance of time-sharing can be gained by considering two types of situations in which time-sharing has so far been found: displacement activities and sequences involving alternating activities.

1. Displacement Activities

McFarland (1966b) thought of the functional significance of displacement activities as a by-product of a mechanism which enables animals to break away from a specific course of action, when progress becomes slow. This conclusion is echoed in two recent studies in the psychological literature. Staddon and Simmelhag (1971) recorded in detail the behavior of pigeons in Skinner boxes, particularly the interim activities that occur between rewards. These activities are similar to displacement activities in their apparent irrelevance, their association with situations in which a strong drive is blocked, their modifiability by relevant motivational factors. Staddon and Simmelhag interpret such activities in terms of behavioral variation, analogous to genetic variation in evolutionary theory. In a recent theoretical review, Falk (1971) takes a similar approach. Falk notes that in situations where rate of consummatory activity is limited, adjunctive behavior occurs, which is similar to displacement activity. This adjunctive behavior may take extreme forms, such as excessive drinking or extreme aggression to a conspecific. This type of behavior, and psychogenic polydipsia in particular, has been subject to intense research during the past 10 years, and remains a persistent problem. After reviewing the innumerable blind alleys that have been explored in this research, Falk concludes by quoting Armstrong (1950), "a species which is able to modify its behavior to suit changed circumstances by means of displacements, rather than by the evolution of ad hoc modifications starting from scratch, will have an advantage over other species."

The idea of displacement activities as a kind of insurance against hard times has probably been overdone. It is difficult to believe that widespread disinhibition, acting as a license for a variety of undisciplined activities, can provide a satisfactory means of organizing behavior. An alternative hypothesis is provided by the possibility that a dominant activity may both disinhibit and activate a subsequent activity. This does not in any way invalidate the disinhibition concept *per se*. Fentress (1968) obtained some evidence that displacement grooming in voles is not only disinhibited but may also be activated by a variety of situational factors. A further example is provided by the phenomenon of food-associated drinking. Recent work shows that, in both rats and doves, feeding behavior has a short-term augmenting effect on the tendency to drink, but that such drinking is a disinhibited activity (Kissileff, 1969; McFarland, 1969). Food-associated drinking appears to be causally independent of the state of thirst, but by allowing time to drink in association with meals, the animal is able to anticipate the thirst-inducing consequences of food intake, and thus achieve a more accurate regulation. Further examples have appeared recently in the displacement literature. Wilz (1970a,b) provides evidence that the so-called displacement activities performed by the three-spined stickleback at the nest are not merely an epiphenomenon of internal events, but are instrumental in effecting a motivational switch to a state favorable to sexual activity. Wilz thus advocates a self-regulation of motivational state, similar to that previously suggested by Delius (1967). The work of Feekes (1971) on displacement ground-pecking in jungle fowl reaches a similar conclusion.

A theory involving disinhibition as the sole determinant of displacement activities, such as that proposed by McFarland (1966b), involves the basic assumption that the probability of any particular activity appearing as a displacement is unconstrained, and therefore that displacement activities cannot serve specific functions, but only a general type of function.

While purely unconstrained disinhibition may occur in some cases, and may indeed be important in providing opportunity for animals to adapt to new environmental circumstances, there will often be constraints upon the type of behavior that is disinhibited. Such constraints may arise as a consequence of the environmental situation in which the disinhibition normally occurs. For example, working with the jungle fowl (*Gallus gallus spadiceus*), Feekes (1971) found that factors normally controlling feeding also control displacement ground-pecking in agonistic situations. The amount of ground-pecking was influenced by the presence of grain, or other stimuli eliciting pecking, such as sand particles. As such particles would normally be present in a natural situation, we might expect ground-pecking normally to have high priority as a displacement activity. Constraints may also arise as a consequence of internal factors. For example, McFarland (1965) found that displacement feeding in thirsty doves was correlated with the degree of thirst-induced hunger, and concluded that displacement feeding is likely to have high priority in thirsty animals. In general, a satisfactory explanation for the occurrence of particular displacement activities can only result from intensive investigation of the particular circumstances involved.

Apart from the question of constraints involved in the elicitation of displacement activities, there are also questions concerning the consequences of displacement activities. In particular, do displacement activities directly influence other aspects of the motivational state? Feekes (1971) found that ground-pecking in jungle fowl was eliminated in the absence of sand particles, and the absence of such pecking was correlated with changes in agonistic behavior. Subsequent experiments suggested that the performance of ground pecks decreases alarm and increases aggressive behavior and thus functions in the resolution of approach-avoidance conflict typical of agonistic situations. Similarly, Wilz (1970a,b) suggested that the performance of displacement activities at the nest increased subsequent sexual behavior in sticklebacks.

In conclusion, it may be that displacement activities serve an opportunistic function, but it seems that they may also influence motivational state, and thus help to resolve conflict situations. In addition, there is the well-documented view that displacement activities may become ritualized in the course of evolution and come to serve a communication function (Tinbergen, 1952).

2. *Alternating Activities*

Behavioral sequences involving repeated alternation of two activities can be classified in three ways, as illustrated in Fig. 1. Of these classes, competition and chaining embody no special principles of organization other than that of behavioral inhibition. Time-sharing, however, appears to involve an additional principle of temporal organization. Time-sharing implies the existence of an underlying control system about which we know nothing. Moreover we can only speculate about the functional significance of such control.

In the case of stickleback courtship, it seems to me that successful courtship involves not only inducing a female to lay eggs in the nest, but also maintaining the nest in good condition to receive the eggs. It would be maladaptive if, having led a female to the nest, the male found it destroyed. Perhaps such a situation could be avoided if the male could check the nest from time to time during courtship. To do this he would have to be able to prevent the female from following to eagerly.

The work of Wilz suggests that, under certain conditions, continuous following by the female decreases rather than increases the likelihood of male sexual behavior. In response to female approach an aggressive male will give a *dorsal pricking* display, which appears to induce the female to adopt a holding posture. The male then visits the nest. According to Wilz, the male uses this breathing space to perform activities at the nest which are necessary to induce a shift from predominant aggressive to sexual motivation. If prevented from indulging in these activities, the shift in motivational balance does not occur.

In the stickleback experiment described earlier, I took the opportunity of disturbing the nest by dropping small bits of weed onto it. The temporal pattern of the male behavior was not affected by this, but the type of behavior at the nest was affected. The male spent his nest visits in removing excess material, instead of fanning or creeping through. This clearing-up operation was interrupted by returns to the female, and resumed on the next visit to the nest.

Suppose that a function underlying the organization of courtship was the maintenance of the nest. The male stickleback would be like a juggler trying to keep two balls in the air. The female could not be left for too long while the male attended to the nest, nor could the nest be left for too long, while he attended to the female. In other words, some kind of time-sharing would be necessary. A corollary of this type of argument is that time-sharing between *two alternating* activities suggests that a double survival value is involved, each "goal" being time-dependent. McFarland (1971) has suggested a similar type of explanation for the existence of time-sharing in the behavioral restoration of physiological imbalance, such as the feeding-drinking experiment discussed above.

The existence of displacements in motivational state, corresponding to physiological imbalances, represents a potential danger to the animal in that the physiological consequences of the value of any displacement exceeding a certain value may be lethal. The probability P_x of such a displacement exceeding a boundary value increases as the displacement x increases, because the opportunity for the animal to reduce x by appropriate behavior is lessened. The probability P_b of any one of a number of displacements exceeding a boundary value is given by the equation

$$P_b = P_x + P_a - P_x P_a,$$

where P_x is the probability of a particular displacement exceeding a boundary and P_a the probability of another displacement doing so. Suppose that an animal has the opportunity to reduce P_x by a given amount in a given time, or to reduce both P_x and P_a by half that amount in the same time. The latter course will always result in a lower P_b. Thus

$$(P_x - \tfrac{1}{2}P_r) + (P_a - \tfrac{1}{2}P_r) - (P_x - \tfrac{1}{2}P_r)(P_a - \tfrac{1}{2}P_r)$$
$$< (P_x - P_r) + P_a - (P_x - P_r)P_a$$

where P_r is the probability reduction that is possible in the time available. In general, when an animal time-shares between activities that reduce the probability of a boundary being exceeded an advantage is gained in terms of the overall probability of a lethal boundary being exceeded.

Although these suggestions concerning the functional significance of time-sharing may have a certain validity, a great deal of research is required before they can be claimed to have veracity. Even with the small amount of work that has been done so far, there are a number of problems. For example, McFarland (1970) found that satiation curves for feeding in doves were the same whether or not water was available, even in animals which had no experience of drinking in the experimental situation. It would appear that the opportunity for time-sharing, rather than sharing itself, may be the important factor.

References

Armstrong, E. A. 1950. The nature and function of displacement activities. *Symp. Soc. Exp. Biol.* **4**, 361–384.

Baerends, G. P., Brouwer, R., and Waterbolk, H. T. 1955. Ethological studies on *Lebistes reticulatus* (Peters): I. An analysis of the male courtship pattern. *Behaviour* **8**, 249–334.

Cane, V. 1961. Some ways of describing behaviour. *In* "Current Problems in Animal Behaviour" (W. H. Thorpe and O. L. Zangwill, eds.), pp. 361–388. Cambridge Univ. Press, London and New York.

Cicala, G. A. 1961. Running speed in rats as a function of drive level and presence or absence of competing response trials. *J. Exp. Psychol.* **62**, 329–334.

Cotton, J. W. 1953. Running time as a function of amount of food deprivation. *J. Exp. Psychol.* **46**, 188–198.

Delius, J. D. 1967. Displacement activities and arousal. *Nature (London)* **214**, 1259–1260.

Falk, J. L. 1969. Conditions producing psychogenic polydipsia in animals. *Ann N.Y. Acad. Sci.* **151**, 569–593.

Falk, J. L. 1971. The nature and determinants of adjunctive behavior. *Physiol. Behav.* **6**, 577–588.

Feekes, F. 1971. "Irrelevant" ground pecking in agonistic situations in burmese red jungle fowl (*Gallus gallus spadiceus*). Ph.D. Thesis, Univ. of Groningen, Groningen.

Fentress, J. C. 1968. Interrupted ongoing behaviour in two species of vole (*Microtus agrestis* and *Clethrionomys brithannicus*). *Anim. Behav.* **16**, 135–153, 154–167.

Fitzsimons, J. T., and Le Magnen, J. 1969. Eating as a regulatory control of drinking in the rat. *J. Comp. Physiol. Psychol.* **67**, 273–283.

Gibb, J. C. B. 1954. Feeding ecology of tits, with notes on treecreeper and goldcrest. *Ibis* **96**, 513–543.

Hinde, R. A. 1970. "Animal Behaviour. A Synthesis of Ethology and Comparative Psychology," 2nd Ed. McGraw-Hill, London.

Hinde, R. A., and Stevenson, J. G. 1969. Sequences of behavior. *In* "Advances in the Study of Behavior" (D. S. Lehrman, R. A. Hinde, and E. Shaw, eds.), Vol. 2, pp. 267–296. Academic Press, New York.

Kissileff, H. R. 1969. Food-associated drinking in the rat. *J. Comp. Physiol. Psychol.* **65**, 284–300.

Logan, F. A. 1964. The free behaviour situation. *In* "Nebraska Symposium on Motivation" (D. Levine, ed.), Vol. 12, pp. 99–129. Univ. of Nebraska Press, Lincoln.

McFarland, D. J. 1965. Hunger, thirst and displacement pecking in the Barbary dove. *Anim. Behav.* **13**, 292–300.

McFarland, D. J. 1966a. The role of attention in the disinhibition of displacement activity. *Quart. J. Exp. Psychol.* **18**, 19–30.

McFarland, D. J. 1966b. On the causal and functional significance of displacement activities. *Z. Tierpsychol.* **23**, 217–235.

McFarland, D. J. 1969. Mechanisms of behavioural disinhibition. *Anim. Behav.* **17**, 238–242.

McFarland, D. J. 1970. Adjunctive behaviour in feeding and drinking situations. *Rev. Comp. Anim.* **4**, 64–73.

McFarland, D. J. 1971. "Feedback Mechanisms in Animal Behaviour." Academic Press, New York.

McFarland, D. J., and L'Angellier, A. B. 1966. Disinhibition of drinking during satiation of feeding behaviour in the Barbary dove. *Anim. Behav.* **14**, 463–467.

McFarland, D. J., and Lloyd, I. 1973. Time-shared feeding and drinking. *Quart. J. Exp. Psychol.,* **25**, 48–61.

Miller, N. E. 1944. Experimental studies of conflict. *In* "Personality and Behaviour Disorders" (J. M. Hunt, ed.), pp. 431–465. Ronald Press, New York.

Rowell, C. H. F. 1961. Displacement activity in the Chaffinch. *Anim. Behav.* **9**, 38–63.

Sevenster, P. A. 1961. A causal analysis of a displacement activity (fanning in *Gasterosteus aculeatus* L.). *Behaviour, Suppl.* **8**.

Sherrington, C. S. 1906. "The Integrative Action of the Nervous System." Yale Univ. Press, New Haven, Connecticut.

Staddon, J. E. R., and Simmelhag, V. L. 1971. The superstition experiment. A reexamination of its implications for the principles of adaptive behavior. *Psychol. Rev.* **78**, 3–43.

Tinbergen, N. 1951. "The Study of Instinct." Oxford Univ. Press (Clarendon), London and New York.

Tinbergen, N. 1952. "Derived' activities: Their causation, biological significance, origin and emanicipation during evolution. *Quart. Rev. Biol.* **27**, 1–32.

van Iersel, J. J., and Bol, A. C. A. 1958. Preening of two term species. A study of displacement activities. *Behaviour* **13**, 1–88.

von Holst, E., and von Saint Paul, U. 1963. On the functional organisation of drives. *Anim. Behav.* **11**, 1–20.

Wiepkema, P. 1961. An ethological analysis of the reproductive behaviour of the bitterling. *Arch. Neer. Zool.* **14**, 103–199.

Wilz, V. J. 1970a. Causal and functional analysis of dorsal pricking and nest activity in the courtship of the three-spined stickleback, *Gasterosteus aculeatus*. *Anim. Behav.* **18**, 115–124.

Wilz, V. J. 1970b. The disinhibition interpretation of the "displacement" activities during courtship in the three-spined stickleback, *Gasterosteus aculeatus*. *Anim. Behav.* **18**, 682–687.

Male-Female Interactions and the Organization of Mammalian Mating Patterns

CAROL DIAKOW[1]

THE ROCKEFELLER UNIVERSITY
NEW YORK, NEW YORK

[1] Present Address: Department of Biology, Adelphi University, Garden City, Long Island, New York.

I. Introduction

Mating behavior in nonhuman mammals is usually characterized by species-specific patterns composed of behavioral elements which have predictable frequencies as well as predictable temporal sequences. These patterns, exhibited by male and female, result in coordination of the mating pair so that insemination can occur. Among the many determinants of the organization of mating patterns are evolutionary adaptations, genetic factors, and physiological interrelationships (i.e., hormonal and neural). In addition, the organization of mating patterns depends on the social experience of each participant in sexual behavior, that is, on the effects of stimulation by other animals.

One objective of this paper is to show how three sources of social experience affect the mating patterns of nonhuman mammals. These are prepuberal social experience, previous mating experience, and experience at the time that mating patterns are being exhibited. There is experimental evidence that social experience prepuberally and previous mating experience postpuberally affect the probability that mating responses will occur, that previous mating experience affects the temporal aspects of the mating pattern as well, and that mating patterns are organized and coordinated by interaction of male and female during sexual encounters. The second objective of this paper is to describe functions of specific sensory modalities in organizing and coordinating male and female patterns. The third objective is to demonstrate how the study of internal processes underlying mating behavior can be enriched by understanding external factors in mating.

Previous reviews of sensory factors in mating behavior have been included in the general accounts of sexual behavior by Beach (1947, 1951) and in those accounts of Schein and Hale (1965) on birds and of Aronson (1965) on lower vertebrates.

In view of Beach's urgings (1965, 1970), some definition of terms as they are used in this paper is in order. Mating behavior refers to the interactions that can be observed when male and female encounter each other in a situation that leads to copulation. This definition includes investigatory behavior, mounting, lordosis, intromission, ejaculation, and any observable postcopulatory reactions. Unless otherwise indicated, the terms copulation and intromission include all the movements associated with penetration of the penis into the vagina, including the accompanying mount. This definition is consistent with the use of these terms in the literature dealing with mating behavior, wherein a mount which includes intromission is identified by the movements accompanying penile insertion and not by observation of penile-vaginal contact. Likewise, the term ejaculation refers to all the movements associated with insemination from the time of initiation of the

mount that includes emission. A successful mating series is one that includes ejaculation.

II. EFFECTS OF EXPERIENCE PREVIOUS TO THE MATING ENCOUNTER

The literature dealing with the effect on mating performance of experience previous to the mating encounter has been reviewed by Beach (1951), Lehrman (1962), and Rosenblatt (1965). The following sections outline studies which elucidate some of the principles which these reviewers have elaborated.

A. PREPUBERAL SOCIAL EXPERIENCE

Restricting contact with other individuals prepuberally affects the mating ability, in adulthood, of male monkeys (Mason, 1960; Harlow, 1965; Missakian, 1969), guinea pigs (Valenstein et al., 1955), rats (Beach, 1942b; Kagan and Beach, 1953; Folman and Drori, 1965; Gerall et al., 1967; Gruendel and Arnold, 1969), and cats (Rosenblatt, 1965). These effects of prepuberal restriction of social experience (often combined with some postpuberal restriction as well) have been either impairment (Table I) or facilitation of mating ability.

There have been reports of both impairment and facilitation of sexual behavior resulting from restricted social experience even within a single species. Among rats, males reared alone were less likely to achieve intromissions than males raised socially either with males only or in heterosexual groups (Folman and Drori, 1965; Gerall et al., 1967; Hard and Larsson, 1968a). On the other hand it has been reported (Beach, 1942b; Kagan and Beach, 1953; Gerall et al., 1967) that early social isolation facilitates the appearance of intromission. Efforts to reconcile these opposing results have focused on the appearance of competing responses (e.g., play behavior) in socially raised males that are absent in the isolate reared animals (Kagan and Beach, 1953; Gerall et al., 1967). However, Hard and Larsson (1968a) found direct evidence of the effects of social isolation in the disoriented mounting that was prevalent in the isolates during mating tests with estrous females. During their first experience with estrous females, socially reared males displayed disoriented mounting in only 8 out of 18 cases while all 17 isolate reared males exhibited disoriented mounting.

Among male monkeys that have been reared alone there is persistent disoriented mounting of females. In addition, other social relations with normally raised females were unusual in that the females "cowered and gri-

TABLE I

IMPAIRMENT OF THE BEHAVIOR OF ADULT MALE GUINEA PIGS BY SOCIAL RESTRICTION DURING REARING[a,b]

	Number	Lower measures[c]		Mountings		Intromissions		Ejaculations		Average scores
		Average per animal	Percent displaying	Average per animal	Percent displaying	Average per animal	Percent displaying	Average per animal	Percent displaying	
Isolated	10	120.8	100	10.6	30	6.5	30	1.5	30	4.9
Social situation[d]	10	52.9	100	14.3	100	19.6	100	5.3	100	8.1

[a] From Valenstein et al. (1955). Copyright 1955 by the American Psychological Association. Reprinted by permission.
[b] Data obtained from 7 tests of each male during isolation, days 77–120.
[c] The lower measures score is the frequency of sniffing, nuzzling, and abortive mounts.
[d] The socially reared males were raised with females.

maced" in the presence of these males more than they did in the presence of other males (Mason, 1960).

B. ADULT MATING EXPERIENCE

A previously unmated animal mates differently from one which has mated before, even though the motor components of the pattern appear the same. Larsson (1959a) compared sexually experienced male rats with naïve males of the same age in a test situation in which they were permitted to ejaculate four times. The interval between an ejaculation and the next intromission was longer in the previously unmated males, otherwise their behavior seemed identical to that of previously mated males. Dewsbury (1969) found, in addition, that it took longer for naïve males to ejaculate once they had achieved the first intromission and that there was a tendency for them to intromit more frequently before ejaculating. Male mice with less mating experience showed more disorientation while mounting than those with more experience (McGill, 1962); however, other behavior patterns accompanying mating, such as pursuing, genital grooming, sniffing, and so forth, were not different in the naïve and previously mated male rats (Mosig and Dewsbury, 1970).

Although the experimental evidence is limited, sexual experience apparently affects the mating behavior of females also. Female cats were more receptive after experiencing a successful intromission than they were before this intromission (Whalen, 1963a). Five females, acquired by the laboratory prepuberally and not allowed mating experience before the experiment were given 1, 2, 3, 4, or 6 tests during each of which they were caged with vigorous, sexually experienced males for 30 minutes. There were few mounts and no intromissions because the females "repulsed the approaches of the male until the test ended." The experimenter held the female until the male intromitted, but on subsequent tests did not interfere in the interactions of the male and female. On the test following this manipulation, all females allowed the males to neck grip and mount. By the third test, all pairs had at least one spontaneous intromission. Intromission frequency increased and intromission latency decreased with continued testing. The increased receptivity after mating was not due to changes in gonadal hormone levels because all the females were spayed and were receiving exogenous estrogen.

The magnitude of the effect of mating experience on the performance of mating behavior is seen in the results of experiments on the persistence of mating after castration in males. Males of several species eventually lose the ability to mate after castration, but sexually experienced male cats (Rosenblatt and Aronson, 1958; Rosenblatt, 1965) and hamsters (Bunnell and Kimmel, 1965) maintained their ability to mate longer than less experienced males. This was not true for rats (Stone, 1927; Rabedeau and Whalen, 1959).

III. Effects of Experience during the Mating Encounter

A mating pattern is an organized sequence of events, and its pacing and coordination are determined by actual performance of the elements of the mating pattern as well as by the nature of the mating partner. In some senses these two are not mutually exclusive, but for clarity of presentation the first section below examines how the performance of each component of the mating pattern of the male affects that and other components, leaving out those effects which are known to emanate from the female and those which are produced in her. Then, the second section will consider the latter.

A. Effects of the Performance of Elements of the Mating Pattern

The mating pattern of male rats consists of a series of mounts and intromissions followed by an ejaculation; this series may be repeated. The temporal aspects of this sequence have been described extensively (Stone and Ferguson, 1940; Peirce and Nuttall, 1961a; Young, 1961; Bermant, 1965; Dewsbury, 1967; Sachs and Barfield, 1970). Carlsson and Larsson (1962) reported that the first ejaculation is preceded, on the average, by 9 intromissions separated from one another by a mean interval of 0.4 minutes. Genital licking usually follows each intromission. There are changes with continued series of mounts to ejaculation and these have been well documented (Table II). The male resumes mounting within 400 seconds of the first ejaculation (Beach and Jordan, 1956; Dewsbury, 1968). Fewer intromissions and less

TABLE II

Changes in Performance Associated with Successive Ejaculations[a,b]

Serial number of ejaculation	Intromissions to produce ejaculation	Seconds to achieve ejaculation	Seconds to recover and resume mating
1	10.64	450	324
2	6.00	216	395
3	5.73	198	468
4	5.09	150	495
5	5.09	210	597
6	4.10	132	818

[a] From Beach and Jordan (1956).
[b] Data obtained from laboratory rats.

time precede the next ejaculation, but it takes more time to recover from each succeeding ejaculation (Beach and Jordan, 1956), and this is also true for hamsters (Beach and Rabedeau, 1959). Male rats are usually considered exhausted or satiated if 30 minutes pass without resumption of mounting. With this criterion, male rats are "exhausted" after a median of 6.9 ejaculations (Beach and Jordan, 1956) ; ejaculation latency and mean interintromission interval increase as male rats approach satiety (Dewsbury, 1968). The mating pattern of the male rat is fairly predictable in terms of numbers of components and the intervals between them. Performance of each component contributes to this regularity as is shown by a body of experiments which changed the characteristics of the pattern by manipulation of the numbers of and the length of the intervals between mounts, intromissions, and ejaculations.

1. Mounts

Mounts without intromission facilitate ejaculation according to Hard and Larsson (1968b). Each male rat was given a 40-minute test with a female whose vagina had been sewn closed; then, he was put with a normal female. In the pretest, the males mounted from 9 to 78 times without intromitting. In the retest, the number of mounts, the number of intromissions, and the time required for ejaculation were all lower than normal.

2. Intromissions

The interval between intromissions determines the number of intromissions required to achieve ejaculation. By lengthening the interintromission interval, Larsson (1959c) reduced the number of intromissions before ejaculation. The interintromission interval was lengthened by removing the male from the female after every intromission, and the effect was strongest with an enforced interval of 2 minutes (Bermant et al., 1969). Long intervals enforced between intromissions have also decreased mount frequency and postejaculatory intervals (Bermant, 1964) and have increased the duration of intromissions (Peirce and Nuttall, 1961a).

These effects have been interpreted as meaning that successive intromissions build up "excitement" within the central nervous system until enough excitement is generated that a threshold is reached and ejaculation occurs (Carlsson and Larsson, 1962; Bermant et al., 1969; Beach and Whalen, 1959a; Larsson, 1961a; Bermant, 1967). The physiological basis for this excitement is generally thought to be a neural change, but it has not been identified. Presumably the excitatory effect of a single intromission reaches a maximum well after the next intromission normally occurs, therefore, fewer intromissions are required for reaching this ejaculatory threshold as the interintromission interval is lengthened. Empirical data have suggested to Larsson (1960) that the excitatory effect of intromission can last for as long as 120 minutes. This suggestion was based on the finding that allowing one or more intromissions 20–120 minutes before an uninterrupted mating test reduced the number of intromissions before ejaculation.

3. Ejaculation

There is usually a period of sexual inactivity before a male resumes mating after an ejaculation. The length of this interval between an ejaculation and the next series of mounts to ejaculation affects the behavior of the male during the second encounter. When this postejaculatory interval was increased, the number of intromissions before the second ejaculation was increased (Larsson, 1958; Beach and Whalen, 1959b), but the intervals between intromissions were shortened resulting in shorter ejaculation latency (Larsson, 1959b; Dewsbury and Bolce, 1970; Beach and Whalen, 1959b). A postejaculatory rest period of 7 hours, which was enforced by separating the male and female, reduced the mount latency of the next series to a minimum. These effects waned after different lengths of time. If the rest exceeded 5 hours, there was no reduction of interintromission interval (Beach and Whalen, 1959b), and with rest periods 90 minutes or longer, there was no reduction in ejaculatory latency (Larsson, 1961b). This suggests that during the interval after ejaculation there are internal events caused by the performance of the preceding series of mounts, intromissions, and ejaculation that have effects on the next series. The nature of these events is unknown.

The effects of rest between mating bouts on male sex behavior may not have been due to "rest" at all, but may have been due in part to the use of an experimental design involving repeated mating with the same female. Repeated mating may have affected the female, and her altered behavior may have produced changes in the males' mating patterns in the second series. Consistent with this interpretation is the demonstration by Dewsbury and Bolce (1968) that there are no effects of rest when the male is given a new female for the second series of mounts to ejaculation.

The extent of these effects depends on the individual male, since the facilitatory effect of the prolonged rest after ejaculation (i.e., shortening of latency to next ejaculation) lasted longer in males with long ejaculation latencies and interintromission intervals than in males for whom these measures were short (Dewsbury and Bolce, 1968).

Seminal discharge produced by electro-ejaculation before a male was placed with a female had no effect on the mating pattern of male rats (Arvidsson and Larsson, 1967). Emission, itself, therefore, is not the cause of the usual changes in the temporal aspects of the mating sequence which occurs after a series of mounts with ejaculation.

4. Genital Grooming

After intromission, the male rat licks its penis. Hart and Haugen (1971a) investigated the role of this grooming in the mating pattern, and found that the prevention of genital licking had no effect on pacing of the components of the pattern. Mount and intromission frequency, intromission and ejaculation latency, and the postejaculatory intervals were the same in animals with full collars which prevented genital grooming and in animals with partial collars that were less prohibitive. However, in males with full collars, erections were observed after intromission and the penis protruded from the sheath until ejaculation. After ejaculation, seminal fluid coagulated at the penis tip. Thus genital grooming serves a cleaning function and aids detumescence of the penis rather than pacing the gross movements during mating.

B. INFLUENCE OF THE MATING PARTNER

1. Female Role in Pacing the Mating Pattern

Female rats approach and stay with male rats, and these tendencies are enhanced when they are in heat. This was demonstrated clearly by Meyerson and Lindstrom (1971) who showed that estrogen-treated female rats crossed an electrified grid more frequently and sustained a higher current to do so when the incentive in the goal box was a male rather than another female.

In contrast to a tendency to approach a possible sexual partner, female rats actively avoid contact with males after sexual encounters. When placed in a situation in which they could avoid the males with whom they were copulating by jumping into an escape box, female rats spent more time away from the males after mounts with intromission than after simple mounts, and still more time away from the males after ejaculation (Peirce and Nuttall, 1961b). To study the same question, Bermant (1961) trained female rats to press a lever for contact with a male. When the female pressed the bar, the experimenter placed a male in her box, observed the nature of the contact with the male, and removed the male. Bermant observed that the length of the interval before the next bar press depended on the nature of the contact. After mounts only it was always less than 0.5 minute, after intromission it was as little as 0.4 minute, but not more than 1.5 minutes, and after ejaculation it was always between 0.5 minute and 2.5 minutes.

Female cats are also unreceptive for a time after sexual contact. It is more obvious in cats than in rats, since the female cat often throws the mounted male off her back and hisses and strikes at him after intromission.

She will not allow another mount for some minutes afterwards as noted by Whalen (1963b), Diakow (1971a), and others.

2. Male Reaction to Females

The observed timing of the male rat's activities during mating depends on whether a male has access to more than one female at a time, on the number of females with which he is paired successively, and on the recent sexual activity of the female with whom he is paired.

When several females were accessible to the male rat simultaneously, the male mounted and intromitted more frequently before the first ejaculation and took longer to ejaculate than when there was only one female present. However, there was no difference in the number of ejaculations achieved before exhaustion compared to when the males had access to only one female (Tiefer, 1969). ("Exhaustion" is a term used to denote varying criteria for cessation of active mating by the males. In this case, the exhaustion criterion was 30 minutes with no intromission or 60 minutes with no ejaculation.)

When males mated to exhaustion with only one female, they were re-aroused by a new female. Fisher (1962) placed male rats individually with estrous females until 45 minutes passed with no mounts. With this criterion of exhaustion, the median numbers of ejaculations and intromissions were 7 and 40.5 respectively. If the females were replaced with recently unmated females after 15 minutes with no mounts, the sexual capacities of the males were greater, that is, the median numbers of ejaculations and intromissions were raised to 13 and 85.5, respectively. This facilitation of mating behavior by a new female has been called the "Coolidge Effect" and has been demon-strated in male guinea pigs (Grunt and Young, 1952) and rams (Beamer et al., 1969), as well as in other species (Fig. 1). Rearousal of sexual interest by new mating partners has not occurred in males under all circumstances: it is more likely to occur in the early part of the males' refractory period than later on (Bermant et al., 1968). In addition, the new female must actually mate with the male, if the male is to be rearoused. When new fe-males were merely caged within sight, sound, and smell of the male, the male's performance with his original mating partner did not improve (Cherney and Bermant, 1970).

The stimulatory effect of the female depends on her recent sexual activity. Females unmated for several days before testing as well as recently mated females reactivated male rats (Hsiao, 1965) but the former were more effec-tive than the latter (Fowler and Whalen, 1961; Wilson et al., 1963).

3. Female Preferences for Individual Males

A female's preference for a particular mating partner is a factor in deter-mining the interactions of the mating pair. Elaborate studies of this phe-

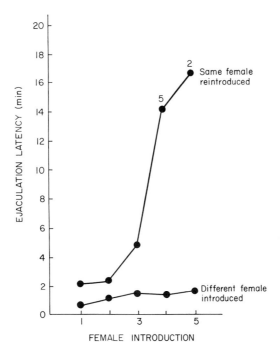

FIG. 1. These data show that a novel female rearouses a previously sexually satiated male ram. Six males were allowed copulation each with a single female for 1 hour (session 1); 6 days later, a different estrous ewe was introduced 5 minutes after each ejaculation (session 2; lower curve); and again 6 days later, the same female was removed after ejaculation and reintroduced four times (session 3, upper curve). (From Beamer *et al.*, 1969.)

nomenon in dogs have been presented (Beach and LeBoeuf, 1967; LeBoeuf, 1967). They have shown that individual female beagles exhibited different levels of acceptance or rejection behavior toward different males. A female's rejection of a male appeared to lower the tendency of that male to mount. The effects of the female's rejection or acceptance behavior on the outcome of a pairing (i.e., on the probability of the occurrence of a lock) were difficult to evaluate, however. There were multiple causes for this: for example, one male was allowed to mount, but he rarely achieved intromission; another male was rejected frequently, but if he mounted there was a high probability of an intromission and lock.

4. *Effect of the Male's Behavior on the Probability of Eliciting the Female's Responses*

When female rats are mounted, they may facilitate penile intromission by showing lordosis, a standing posture involving dorsoflexion of the vertebral column; or, they may hamper mounting by rolling over on their backs or kicking with their hind legs. Repeated mounts with or without intromission affect these responses of the female rat. In one study (Hardy and De-Bold, 1971), twenty female rats were mounted 50 times each by males who could not intromit because adhesive tape covered the perineal regions of the females. The effects of these repeated mounts without intromission were that the females were less likely to show lordsis during subsequent mounts than females which had not received the pretest mounts. The pretest mounts also decreased rejection of the males and decreased the intensity of the lordoses which were shown.

Another study (Hardy and DeBold, 1972) concerned the effects of mounts with intromission. By comparing the behavior of females with adhesive tape covering their perineal regions and females without such tape, they found that females experiencing the full intromission pattern of the male for 10 or more mounts were less likely to show lordosis, showed less intense lordosis, and were more likely to reject the male by rolling over during those mounts.

IV. SPECIFIC AFFERENT ROUTES

The effects of experience on mating reviewed in the previous sections show that experience provides information useful in determining adult sexual performance. This section will examine how the mating animals sense the environment, that is, it will explore specific afferent pathways that function in guiding sexual behavior or that subserve some of the functions of experience examined above.

A. TOUCH

1. Body Areas Other than the Genitalia

a. *Females.* Tactile stimulation of the back, flanks, and perineum of females of many species causes them to assume a mating posture when they are in estrus. Manual stimulation of the back or perineum elicits lordosis in female hamsters (Pfaff *et al.*, 1972); rubbing the back and/or perineum elicits lordosis in cats (Diakow, 1971a) as does manual stimulation of the back, flanks, and perineum of female rats (Ball, 1937; Blandau *et al.*, 1941; Adler and Bell, 1969; Gerall and McCrady, 1970). In the female rat this stimulation consists of placing the palm of the hand on the back, placing the thumb and fourth and fifth fingers on the flanks in front of the hind legs, and wrapping the index and middle fingers around the iliac crests on either side of the tail. Squeezing the animal between these fingers and the palm of the hand elicits the rump and head elevation and lumbar depression which characterizes lordosis in rats. The assumption that back, flank, and perineum stimulation causes lordosis is also supported by film analysis of interactions of male and female rats during mounting (Diakow, 1974). This film analysis showed that head elevation accompanying lordosis always began after the male placed his forepaws on the female's back and began to mount. The mean latency from time of back contact to the start of elevation of the female's head was 0.53 second (range = 0.22–0.96) for mounts with lordosis only, 0.68 second (0.23–2.23) for mounts with lordosis and intromission, and 0.76 second (0.27–1.05) for mounts with lordosis and ejaculation.

b. Males. Male rats mate after denervation of the ventral part of the body (Stone, 1923) and of the mouth and lips (Beach, 1942a). Because detailed qualitative and quantitative description of the way that these males mounted after desensitization was not presented it is difficult to make a statement about the role of these stimuli in guiding the male's behavior.

2. Genital Region

a. Females. There are many studies which have shown that females assume a receptive posture despite reduction or absence of genital sensation. After estrogen treatment, a normal mating pattern was exhibited by a female rat with congenital absence of ovaries, uterine horns, and an unruptured vagina: this seems to preclude the necessity of stimulation from the uterus and vagina for initiation of lordosis (Beach, 1945). In some experiments performed to study reproductive processes other than mating, it was also noticed that the females mated despite the absence of genital sensations. Application of Novocain to the vaginal and vulval region (Fee and Parkes, 1930) and sympathectomy (Labate, 1940) of the female rabbit did not inhibit mating. Female rats mated even though the cervix (Krehbiel, 1948) or parts of the abdominal sympathetic nervous system (Bacq, 1931a) were removed.

Studies designed to determine the role of genital sensation in female mating behavior have incidentally confirmed that desensitized females are capable of mating. Since in these studies, genital organs have often been denervated, if not removed, the afferent neural pathways from the genitalia are shown in Fig. 2 for aid in following the discussion. Afferent sensations may travel from the clitoris via the pudendal nerve and enter the spinal cord through the second and third sacral dorsal roots. Electrophysiological studies have shown that the pudendal nerve carries tactile stimulation from the clitoris and vulva of the female cat (Diakow and K. Cooper, 1966 unpublished) and female rat (Komisaruk *et al.,* 1972). Afferents from the vagina and uterus travel via both the pelvic and the hypogastric nerves, although the latter were shown by electrophysiological methods to carry more of them (Abrahams and Teare, 1969).

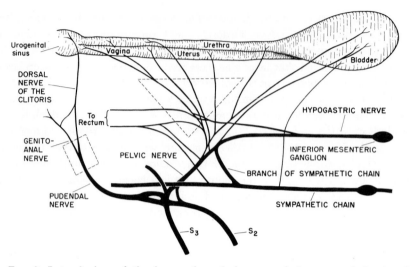

FIG. 2. Lateral view of the innervation of the urogenital organs of the female cat. There are four main nerves of the outer genitalia of the cat. The *pudendal nerve,* innervating the clitoris and vulva, the *pelvic nerve,* the *hypogastric nerve,* and branches from the *sympathetic chain* which innervate the vagina and the body of the uterus. The last three nerves join in the pelvic plexus before distributing to these organs. The dotted lines enclose the nerves cut to desensitize the various areas shown. (From Diakow, 1969.)

Brooks (1937) desensitized the genitalia of female rabbits in various ways (transection or extirpation of the spinal cord in the lumbar or sacral regions, complete abdominal sympathectomy, hysterectomy, removal of the proximal half of the vagina) and found that these females continued to mate. Bard (1935, 1939, 1940) showed that these operations had no effect on ability of cats to show lordosis. Likewise, bilateral removal of the pudendal nerve which desensitized the clitoris and reduced the sensitivity of the vulva had no effect on the mating pattern of cats (Diakow and Aronson, 1967; Diakow, 1971a). Ball (1934a), and later Kaufman (1953), removed the uterus, cervix, vagina, and external opening of the vagina from rats, and both found that nearly all of these females exhibited darting and lordosis characteristic of the species. Similarly, after hypogastric nerve section, guinea pigs mated (Donovan and Traczyk, 1965).

Closer analysis of these early studies, and evidence presented by recent investigations, show that although genital impulses may not play an indispensable role in initiation of mating activity in the female, they are important for her normal responses. Ball (1934a) noted a lower degree of excitability after several matings among females without uteri and vaginas, and Bard (1935) observed failure of response to manual stimulation of the vulva in one female cat from which the sacral portion of the spinal cord was removed. Bilateral pelvic nerve section led to failure of lordosis in response to mechanical stimulation of the vagina and cervix in rats (Carlson and DeFeo, 1965), and in female guinea pigs pelvic neurectomy led to the absence of mating (Donovan and Traczyk, 1965).

What was required to detect the role of genital stimulation, were more detailed observations of normal behavior against which to compare the behavior of desensitized females. The question then could become: "How do females mate after genital desensitization?" rather than "Do females mate after genital desensitization?" The latter questions the role of sensory stimulation in sexual arousal; the former questions the role of sensory stimulation in mating performance. There will be further discussion (Section V,A,2) of the consequences of these two approaches to understanding the physiological bases of behavior.

In cats, vaginal stimulation is responsible for the copulation cry and rubbing and rolling components of the afterreaction which occurs after the female has received an intromission. This was shown in a study in which the vagina and body of the uterus were denervated in ten female cats by bilateral removal of the caudomedial part of the pelvic plexus (see dotted triangle on Fig. 2). The postoperative response to intromission showed the effects of the denervation, directly in the behavior of the female, and indirectly, through effects on the male: Although their preoperative mating patterns were the same as those of eight sham-operated females, postoperatively, the denervated females emitted fewer copulatory cries and rolled less than usual after being mounted. The males' mounts were longer on the denervated females (Diakow and Aronson, 1968; Diakow, 1971a) (Table III).

Cervical stimulation of female rats prolongs lordosis after intromissions and immobilizes the female after ejaculation (Diakow, 1970). After denervation of the cervix by bilateral ablation of the pelvic nerves (possibly also causing reduced vaginal sensation), females' responses to forelimb and some genital movements of the males were compared to those of sham-operated females. The denervated females began to lower their heads 10 milliseconds (this figure is the mean of means) *before* perineal contact was lost following intromission; and they began to move their forepaws 1230 milliseconds *before* the males dismounted following ejaculation. In contrast, sham-operated females lowered their heads from the extreme lordosis position 230 milliseconds *after* perineal contact was lost following intromission; and these females began to move their forepaws 2860 milliseconds *after* dismount following ejaculation. These effects might have been interpreted by Ball (1934a) as a "tendency to lower degree of excitability" in rats without uteri and vaginas.

There are indications of another role for genital sensations in females,

TABLE III

EFFECT OF DENERVATION OF THE UTERUS AND VAGINA ON THE
MATING PATTERN OF FEMALE CATS[a,b]

	Group	
Behavior	Sham operates (mean of subject means)	Denervated females (mean of subject means)
Mount duration (minutes)		
Preoperative	1.4	2.2 |1.
Postoperative	1.2 ——————————— 3.6	
Percentage of mounts ending with a copulatory cry		
Preoperative	84.3	81.9 |
Postoperative	82.1 ——————————— 29.2	
Percentage of time after mounting that female rubbed and rollied		
Preoperative	20.1	13.0 |
Postoperative	16.5 ——————————— 4.1	

[a] From Diakow (1971a).

[b] Horizontal and vertical lines connect means that are significantly different at the $p = 0.05$ level. Mann-Whitney U test was used for between-group comparisons and the Wilcoxon rank sum test for within-group comparisons.

and this is the mediation of the female's refractory period after sexual contact. The refractory period of female cats is very striking: during intromission the female usually turns on the male and throws him off while she hisses and paws at him. This behavior was much reduced in females with denervated genital tracts (Diakow, 1971a). Preoperatively the percent of mounts ending with the female's throwing off the male was 42.2% but it was only 22.7% after denervation. The refractory period of female rats as discussed earlier is probably also mediated by various genital stimuli. Experiments by Bermant and Westbrook (1966) revealed that estrous female rats trained to press a lever for access to males who could mount, intromit, and ejaculate had shorter intervals between lever presses after a local anesthetic had been swabbed on their vaginas and perineal regions.

b. *Males.* The earliest thoughts about a role for genital sensations in sexual behavior concerned males. In the late 1800's, impulses from genital organs were considered a source of "sexual arousal" in males. Tarchanoff, 1887 (cited in Steinach, 1894), claimed to have inhibited amplexus of male frogs by removing or emptying the seminal vesicles, and proposed that tension in the walls of the seminal vesicles caused by swelling and continuous hitting against them by spermatozoa are the stimuli for the release of the amplexus response and for "sexual desire." Steinach (1894) did not agree:

He observed that frogs mated before the seminal vesicles were full and showed that extirpation of the seminal vesicles did not inhibit amplexus. Steinach suggested that bleeding resulting from Tarchanoff's operations inhibited those subjects' mating behavior. Removal of the seminal vesicles also failed to affect the frequency of copulation in rats (Steinach, 1894).

In the 1930's, 1940's, and 1950's the idea that impulses from the male genitalia played no role in maintaining sexual arousal was strengthened. Male rats mated as promptly and frequently after removal of the vasa deferentia as before (Ball, 1934b); the number of mounts exhibited by castrate hamsters decreased equally rapidly after castration whether the males had intact seminal vesicles and prostates or not (Pauker, 1948); and male cats attempted to initiate mating even after removal of the sacral cord and two lumbar segments of the spinal cord (Root and Bard, 1937, 1947).

However, in this period, the idea developed that genital impulses are necessary for the neuromuscular coordination of the complete copulatory pattern. After the sympathetic chains were removed from two adult male rabbits their copulations became abnormal in that they no longer fell on their sides or uttered characteristic cries as is usual at ejaculation (Bacq, 1931b). Beach and Holz (1946) destroyed the integrity of the penis of rats by removing part of the penile bone, and found that the animals did not ejaculate. Since both sexual behavior and the height of penile spines were reduced after castration, Beach and Levinson (1950) proposed that deterioration of the papillae and the consequent loss of sensory input lowers the amount of sexual activity shown by the castrate animal. In the 1960's and 1970's, several investigators strengthened the argument that the sensations arising from the penis during mating coordinate the movements required for intromission and ejaculation and play a role in maintaining sexual arousal. After surgical anesthetization of the penis by bilateral section of the dorsal nerve of the penis, sexually experienced male cats became disoriented during mating and failed to achieve intromission (Cooper and Aronson, 1962; Aronson and Cooper, 1963, 1966, 1968). The same effect was achieved by local anesthesia of the penis (Aronson and Cooper, 1968) and was not due to loss of erection (Cooper et al., 1964). After desensitization of the penis, and consequent loss of intromission, male cats had unusually pronounced seasonal cycles in "sexual arousal." In the fall, mount latency was longer and the sex score (an indicator of total sexual activity) was lower than in the winter (Aronson and Cooper, 1966). Long-term testing revealed a general nonseasonal decline in sexual behavior after 3 or 4 years in these desensitized males (Aronson and Cooper, 1968) (Fig. 3).

Local anesthesia of the penis also makes intromission and ejaculation difficult in rats. When a local anesthetic was applied to the penis of seven sexually experienced male rats, only one male ejaculated; all seven of the control

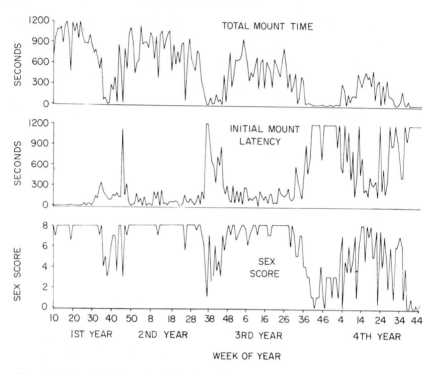

FIG. 3. This illustrates the behavior of male cats after penile desensitization. The sex score is an index of the maximum amount of sexual activity shown by the males of the experiment. A score of 9 indicates that intromission and ejaculation occurred. According to these data, no male intromitted or ejaculated for at least 5 weeks after denervation. Median mount latency increased and total mount time decreased after long-term testing. A seasonal decrement was imposed on this decline. (From Aronson and Cooper, 1968). Copyright 1968 by Indiana University Press, Bloomington. Reprinted by permission of the publisher.

males to whose penises saline had been applied, ejaculated (Carlsson and Larsson, 1964). These effects were similar to those of sectioning the dorsal nerve of the penis in rats (Larsson and Södersten, 1973). If the anesthesia were applied after an ejaculation, the males mounted as readily as controls but required a longer time to achieve the next intromission (Adler and Bermant, 1966).

Monkeys seem also to depend on penile sensations for ejaculation. There was no ejaculation in three out of four male monkeys in which the penis had been desensitized by bilateral removal of the dorsal nerve of the penis. Partial deafferentiation delayed, but did not prevent, ejaculation (Herbert, 1967). However, the precise stimulus conditions that lead to ejaculation have yet to be determined, because it appears that intromission and thrusting alone are not sufficient to lead to ejaculation. Michael and Saayman (1967) tested intact male monkeys with ovariectomized females, but the males did not ejaculate even though they intromitted and thrust as much as would normally lead to ejaculation.

Recent studies dealing with the removal or denervation of the accessory sexual glands, have failed to reveal a role for these organs in the mating pattern of rats (Beach and Wilson, 1963; Larsson and Swedin, 1971). On the other hand, radical sympathectomy, which affects function of more than the genitalia, caused three dogs to mount and ejaculate less than controls. These sympathectomized dogs, however, were able to show the complete repertoire of normal mating activity (Beach, 1969).

B. OLFACTION

1. Effects of Attempts to Induce Anosmia

The earliest experiments designed to test the role of olfaction in sexual behavior involved ablation of the olfactory bulbs, but these were generally unsuccessful in demonstrating any effect. As examples, three male rats mated after transection of the olfactory bulbs and nerves, and six male rabbits mated and successfully inseminated females although five out of six failed to sniff their mating partners (Stone, 1922, 1925). Likewise, a bulbectomized female rabbit mated, was inseminated, and subsequently ovulated (Brooks, 1937).

In more recent studies, as for example those of Heimer and Larsson (1967) and Bermant and Taylor (1969), olfactory bulbectomy resulted in increased ejaculation latencies in naïve male rats. Testosterone was administered on the possibility that bulbectomy altered testes secretion, but it did not counteract the effect of the lesion (Larsson, 1969). The mating patterns of male mice (Rowe and Edwards, 1972) and hamsters (Doty et al., 1971; Murphy and Schneider, 1970) were affected more drastically by bulbectomy: males no longer mounted. In female guinea pigs, olfactory bulb removal decreased the percentage of animals that mated (Donovan and Kopriva, 1965), and a similar effect was demonstrated in rats (Aron et al., 1970). In both of these studies, naturally cycling females were used and matings were not observed, but were determined by the presence of sperm in the vaginal smear. At least in female rats, administration of estrogen increased the percentage of bulbectomized animals that mated, suggesting that bulbectomy exerted its effect on mating in female rats through changes in hormone secretion (Aron et al., 1970). Bulbectomy lowered the frequency with which female mice, treated with hormones, exhibited lordosis in response to mounting (Thompson and Edwards, 1972), and there is one report that bulbectomy facilitates lordosis in female rats (Moss, 1971). These findings show that there are differences between species and between sexes with regard to the general debilitating or facilitating effects of bulbectomy on mating behavior.

While the above studies show that the normal manifestation of the mating pattern depends on the presence of the olfactory bulbs, these studies are extremely difficult to interpret in terms of a role for the sense of olfaction in mating behavior. There are at least two difficulties in attributing the effects of bulbectomy to a deficit in the sense of smell: First, the intimate connection of the olfactory bulb to the limbic system suggests that the olfactory bulbs do more than process olfactory information, and second, the mating test situations used in the above experiments were probably inappropriate for detecting effects of the loss of the sense of smell.

Leonard (1972) has data relating to the first of these difficulties which suggest that the effects of bulbectomy may be due to more than the loss of the ability to detect odors. Similar deficits in social behaviors of hamsters occurred after unilateral and bilateral olfactory bulb damage, and since the animals could still smell after unilateral damage, she concluded that "not all the behavioral effects of olfactory damage can be attributed to anosmia." Moss (1971) and Murphy and Schneider (1970) failed to find effects of unilateral olfactory bulb damage on sexual behavior, whereas they had noticed changes with bilateral bulbectomy. Thompson and Edwards (1972) found a lowering of lordosis probability with unilateral bulbectomies in female mice. Most other studies have excluded this control. It is questionable to attribute the effects of bulbectomy to anosmia until it is shown that peripherally induced anosmia has the same effect as bulbectomy. In male rats, destruction of the nasal mucosa (Larsson, 1971) had the same effect as bulbectomy, but in male mice, in which bulbectomy results in a complete loss of mating behavior, there was no loss of mounting after peripheral anosmia (Rowe and Smith, 1972).

Production of anosmia by preventing air from reaching the nasal mucosa would seem to be a good way of testing the role of the olfactory sense in mating behavior. This procedure had no effect on the mating pattern of male dogs (Hart and Haugen, 1971b), but did prevent monkeys from pressing a lever for access to estrogen-treated females that were attractive to normal males (Michael and Keverne, 1968; Michael, 1971). In the latter studies, nasal plugs were inserted into sexually experienced male monkeys, thereby producing a reversible anosmia, and the males were presented with ovariectomized females that had estrogen swabbed in their vaginas. The vaginal estrogen makes ovariectomized female monkeys very attractive to males, but the females do not act like receptive females. While the plugs were inserted the males did not press the lever to gain access to the females, but they did so after the nasal plugs were removed.

The findings of these studies lead one to suspect that the role of olfaction might be to make it possible for an individual to discriminate a mating partner. In this case, the types of mating tests and observations used in the past to detect effects of bulbectomy may not have revealed the role of olfaction. It is very striking that, in looking for a role for olfaction in mating behavior, few investigators have ever noticed if males sniff the females with which they are paired. Exceptions are Stone (1922, 1925) and Hart and Haugen (1971b). The former found that bulbectomized male rabbits failed to sniff females; the latter noticed that anosmic male dogs did sniff their mating partners. Unfortunately, neither study presented quantified sniffing data.

It has been reported that it takes longer than normal for male rats to achieve ejaculation after bulbectomy. This effect, rather than being based on the specific importance of olfactory stimulation for mating, might be due to an effect on noncopulatory activity. It is easy to imagine that these males, habituated to the mating cage and its odors before the sensory deprivation, might have spent a lot of time exploring what was, to them, a new and unusual environment without odors. This activity could have distracted them from mating and prolonged ejaculation latencies. Along these lines, in cats the latency of the first mount with intromission (i.e., the ejaculation latency) decreased after olfactory bulb ablation because preliminary sniffing and exploring of the test room and of the female was greatly reduced (Cooper and Aronson, 1972).

It would seem, then, that to determine the role of olfaction in the maintenance of normal mating behavior, one should look at the interaction between the male and female and their behavior toward the environment, rather than simply at mounts, intromissions, and lordosis. If an anosmic male does not mount a female, what does he do instead? Putting one male and one female together in a familiar mating environment may be masking the role of olfaction in mediating discrimination of an appropriate mating partner or preference for a particular one. Studies which have presented an animal with a choice of olfactory stimuli and have compared, quantitatively, differential approaches to the stimuli are reviewed in the next section.

2. Effects of Presenting Choices of Olfactory Stimuli

a. *Males.* Olfactory stimuli may assist in the discrimination of appropriate sexual partners. Male rats were trained to respond to either an estrous or a diestrous female odor at the onset of a tone, and the training was successful whether the males were adults with sexual experience or prepuberal castrates (Carr and Caul, 1962).

Associated with this discrimination there is a preference based on olfactory cues for approaching females in mating condition. Michael and Keverne (1968) have reviewed the evidence for this in dogs, horses, bulls, and rams, and they have shown this to be true in monkeys (their experiment was discussed above).

The most complete studies on olfactory control of discrimination of receptive females by males have been done on rodents. In a three-way choice among male, estrous female, and diestrous female odors, adult male mice chose a female in 67 to 87% of the tests (Chanel and Vernet-Maury, 1963). Sexually experienced male rats preferred the odor of an estrous female to that of an anestrous female (Carr et al., 1965), but this preference was not shown by sexually inexperienced males, by castrates, or by sexually sluggish male rats (Carr et al., 1966; Stern, 1970). The development of adult preference for estrous females depends on the presence of androgens neonatally (Stern, 1970; Johnson and Tiefer, 1972), but the preference for estrous females was not evident in immature males unless they were given testosterone propionate (Carr et al., 1970b).

Preferences shown by sexually experienced males for estrous females depend on the identity of the females as well as the nature of sexual experience of the male. Males experienced in mating with several different partners preferred the odor of novel females, but males experienced in mating with only one partner showed no preference for odors of either novel or familiar females (Carr et al., 1970a).

b. *Females.* Female rats use olfactory cues to discriminate between normal and castrate males (Carr and Caul, 1962) and prefer to approach the normal males (Carr et al., 1965). This preference for appropriate sexual partners was shown by female rats in heat whether or not they had mating experience, but was not shown by anestrous females without sexual experience (Carr et al., 1965). In contrast to males, polygamous females preferred to approach the odor of a previous mating partner rather than that of a new one (Carr et al., 1970a). Unmated females preferred to approach the odor of a novel male over that of a familiar male (Krames, 1970). Immature females raised with females preferred the odor of normal males, but immature females raised with males showed no preference for either normal or castrate males (Carr et al., 1970b).

3. Relevant Structures for Olfactory Responses

Because olfactory bulb removals have such a large effect on mating behavior (reviewed above), they are likely to be involved in mediating the olfactory cues utilized in mating behavior. In addition to the olfactory bulbs, there is evidence suggesting that the preoptic area and the lateral hypothalamus may have relevance for mediating olfactory-cued sexual responses. This evidence is the demonstration that these areas contain neurons responsive to urine odors in mice (Scott and Pfaff, 1970) and rats (Pfaff and Pfaffmann, 1969a,b; Pfaff and Gregory, 1971).

4. Sources of Sexually Attractive Odors

Vaginal and preputial gland secretions have been identified under well-controlled laboratory conditions as sources of odors which attract potential sexual partners. The sexual performance of male *Rhesus* with ovariectomized females was improved when vaginal secretions were collected from estrogen-treated females and applied to the sex skin of otherwise untreated ovariectomized females (Michael and Keverne, 1970). Michael *et al.* (1971) have identified a gas chromatograph fraction, which consists of short-chain aliphatic acids, as the active agent in these vaginal secretions. When these aliphatic acids were applied to the sex skin of ovariectomized females, males increased their mounting attempts and actually ejaculated, whereas, the males failed to do so before the material was applied to the females (Fig. 4).

Rats seem to be attracted by odors produced by the preputial glands. Male and female rats, presented with a choice between the odor of the preputial glands of the opposite sex or the odor of a control substance (muscle or liver), preferred the odor of the preputial gland, but showed no preference for the odor of submaxillary or sublingual glands (Orsulak and Gawienowski, 1972).

Mice can make sex discriminations based on preputial gland secretions carried in urine (Bronson and Caroom, 1971). Sexually experienced female mice in various stages of the estrous cycle preferred the odor of male preputial glands or urine from intact males over the odor of urine from males, from whom the preputial gland had been removed, or female urine (Bronson and Caroom, 1971). In contrast, male guinea pigs preferred urine from females over that of males whether or not the urine was taken directly from the bladder or was naturally voided (Beauchamp and Berüter, 1973). This implies that preputial gland secretions are not the attractive agent in the urine of female guinea pigs.

Male rats were shown to be attracted to estrous female urine (Pfaff and Pfaffmann, 1969a). The source of the odoriferous substance in the urine that is the attractant might be the preputial gland as it is in mice, but the appropriate experiments have yet to be performed to demonstrate this. Anatomical evidence obtained by histological study of the genitalia of the rat (Diakow, unpublished observations) indicates that it is very possible for preputial gland secretion to be discharged with urine of female rats. In these animals, the urethra and the ducts of the preputial glands both open within an enclosure formed by the prepuce. It is a prepuce which is visible on

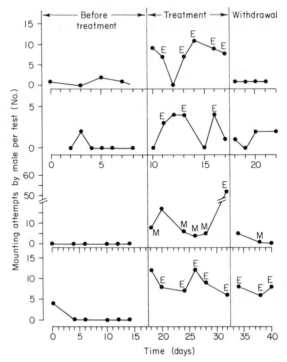

FIG. 4. Sexual stimulation of male rhesus monkeys by components of vaginal secretions fractionated by gas chromatography. In tests with four males each paired with a different ovariectomized (recipient) female, the application to the latters' sexual skin of material collected by trapping from the gas chromatograph resulted in a marked stimulation of the sexual behavior of their male partners. E, ejaculation M, masturbation to ejaculation. Time scale of the lower two pairs tested on alternate days is half that of the upper two pairs tested daily. (From Michael *et al.*, 1971.) Figure and legend copyright 1971 by the American Association for the Advancement of Science.

the outside of the body of the female rat (Greene, 1963) and is commonly, but inaccurately, called the clitoris. This prepuce consists of folds of the skin which enclose the clitoris and the opening of the ducts of the urethra and preputial glands (Diakow, unpublished observations). This anatomical arrangement would enable joint discharge of preputial gland secretion and urine.

C. Vision

Beach (1942a) and Stone (1922, 1923) reported that blind male rats mated. Hard and Larsson (1968c) took detailed quantitative data on the mating of blind males and concurred that their matings were not impaired, but that the intervals between ejaculations were shorter for the blind males. Female rabbits (Brooks, 1937), too, have mated successfully though blind.

D. AUDITION

Male rats (Stone, 1922, 1923) and female rabbits (Brooks, 1937) have mated despite impaired hearing, but ultrasonic sounds have been associated with mating rats and may serve to help coordinate the male and female. These sounds have been described as 50-kilohertz "chirps" which are emitted by males before mounting (Sewell, 1967) and as 22-kilohertz "postejaculatory songs" (Barfield and Geyer, 1972). The latter vocalization occurs after an ejaculation and is accompanied by inactivity of the male. During this time, females tend to stay away from the male. Though a communicatory function of these sounds has not been firmly established, Barfield and Geyer suggest that the postejaculatory song inhibits sexual activity of the female while the male is in his postejaculatory refractory state. Brooks and Banks (1973) have reviewed reports of vocalizations in other rodents and comment on the interpretations of the role of these vocalizations. They have been thought to reduce aggression, or calm down the mating partner, or to indicate willingness to mate; but again, these authors agree that the functions of these sounds have not been established.

V. VALUE OF UNDERSTANDING THE ROLE OF EXTERNAL FACTORS IN BEHAVIOR

We have recently learned a great deal about the neural and hormonal bases for mating behavior by asking questions concerning sensory involvement in this behavior. In particular, we have discovered unsuspected roles for different parts of the nervous system in mating behavior, our understanding of how behavior feeds back to influence the physiology of the behaving organism has grown, and we have been led to ask meaningful questions about interaction of sensory and hormonal determinants of mating behavior.

A. NEW DISCOVERIES CONCERNING THE NEURAL BASES FOR MATING BEHAVIOR

1. Lateral Reticular Nucleus of the Medulla

Because they knew genital stimulation to be important for eliciting mating responses in female cats, Rose and Sutin (1971, 1973) explored the hypothalamus, midbrain, and medulla and found a high concentration of neurons that respond to probing of the urogenital sinus and vagina. They localized this highly responsive region in the lateral reticular nucleus of the medulla. When they lesioned this nucleus, unanesthetized females no longer gave the copulatory cry or rubbing and rolling responses to stimulation, with a glass rod, of the urogenital sinus and vagina. This brain lesion, therefore, had the same effect as the peripheral denervation of the vagina and body of the uterus by removal of the caudomedial part of the pelvic plexus (Diakow, 1971a), and it indicates that the lateral reticular nucleus is a relay in the afferent pathways responsible for the mating pattern.

2. Sensory Thalamus

On the basis of the idea that tactile stimulation is important for the mating posture of female rats, Diakow (1971b) lesioned the ventrobasal complex of the thalamus of the rat. The cells of the ventrobasal nuclei are known to respond to somatosensory stimulation in rats (Emmers, 1965). This lesion prevented the female rats from showing lordosis in response to males as they had done preoperatively. Because previous lesion studies have concentrated on the search for brain areas that mediate sexual drive, arousal, or motivation, or were based on earlier studies that had this goal, most lesions have been placed in areas such as the cortex, hypothalamus, or other areas related to the limbic system which were thought to be important for emotion-related activities. Because of this concern, few experiments have explored the sensory areas of the brain in relation to their function in mating behavior.

3. Midbrain Central Gray

In the rat, Malsbury et al. (1973) have recorded neuronal responses near the midbrain central gray to tactual stimuli of many areas of the body including the back and perineum. This is approximately the same brain region in which electrical stimulation causes rump raising (Pfaff et al., 1972). This indicates that the midbrain might be an area where there is coordination of sensory and motor events indicated as important for lordosis. There is preliminary evidence that lesions in this area reduce lordosis intensity (Diakow, experiment in progress).

4. Spinal Cord

It will be recalled that after desensitization of the cervix, female rats stand still for a shorter time during ejaculatory mounts (Diakow, 1970). It can be concluded, therefore, that stimulation of the cervix normally prolongs standing during mating. On this basis, it might be postulated that stimulation of cervical receptors inhibits limb flexion responses. Along these lines, Komisaruk and Larsson (1971) have found that cervical probing inhibits flexion response to pinching of the legs, and that this inhibition is mediated by the spinal cord.

B. Stimuli Received during Mating Influence the Endocrine
Status of the Individual

It has become clear that sensory stimulation received during mating has important endocrine consequences. In fact, some of the same stimulus pathways mediating behavioral responses during mating also mediate hormonal responses to mating. The following are examples: (1) Afferents from the vagina and body of the uterus, carried through the caudomedial part of the pelvic plexus, are responsible for the normal vocalization and rubbing and rolling by the female which accompanies copulation in cats (Diakow, 1971a). The caudomedial part of the pelvic plexus is also necessary for ovulation in response to artificial stimulation of the distal vagina with a glass rod (Diakow, 1971a). (2) Afferents from the cervix carried through the pelvic nerves are responsible for prolonging lordosis after intromission and standing after ejaculation in female rats (Diakow, 1970). These nerves also are necessary for pseudopregnancy and successful pregnancy (Carlson and DeFeo, 1965; Kollar, 1953).

Other examples of a role for mating stimulation in influencing hormone production are the very interesting findings of Folman and Drori (1969) and Adler (1969). The former showed that mating stimulation increases testosterone production in male rats, and the latter has shown that multiple intromissions induce the hormonal conditions underlying successful implantation.

C. Interactions of Sensory and Hormonal Determinants
of Mating Behavior

A venerable concept, reiterated by Lashley (1938), Beach (1947), and others, is that hormones facilitate the integration of sensory and motor patterns underlying mating. This facilitation has been clearly demonstrated by Komisaruk (1972) and Komisaruk and Diakow (1973) in their study of the hormonal basis of the lordosis posture in female rats. They studied how estrogen affects the probability of eliciting lordosis by manual stimulation consisting of simultaneous application of pressure to the back, flanks, perineum, and cervix of the rat. The use of manual stimulation of the areas appropriate for eliciting lordosis, instead of the use of a sexually active male rat, circumvented confounding of the results by possible hormone influence on mating interactions that occur before mounting and lordosis and limited the observations to the facilitation of the mating posture by exogenous estrogen. Forty-eight percent of 96 ovariectomized females showed lordosis in response to manual stimulation. It was also possible to elicit lordosis in females that were adrenalectomized as well as ovariectomized thus reducing the chances that the lordosis was influenced by extra-gonadal estrogen or progesterone. These findings imply that the motor circuits, and at least some of the sensory circuits, underlying lordosis are present and are able to be activated in the absence of estrogen. Over and above this, estrogen does influence the lordosis response: The percentage of females responding rose in relation to increasing estrogen doses. After receiving a single injection of 0, 0.01, 1.0, or 10.0 μg estradiol benzoate per 100 grams body weight the proportions of rats showing lordosis were: 33.3, 40.0, 90.0, 90.0, and 88.9, respectively.

While estrogen facilitated the stimulus elicitation of the motor pattern there is no way of knowing whether the facilitation was due to an increase in the effective sensory stimulus or greater reactivity of the motor neural substrate. The question of whether or not sex hormones influence behavior by influencing sensory input is an old one (Nissen, 1929; Lashley, 1938), but attempts to answer this question with modern electrophysiological methods are, of course, very recent. In one such study, it was found that while testosterone did not affect activity of single neurons activated by tactile receptors in the penis of the male cat, it did affect the transmission of stimuli from the sensory nerve of the penis to the cortex (Cooper, 1969). Estrogen, on the other hand, seemed to enlarge the receptive field of the pudendal nerve which innervates the clitoris and perineum of the female rat (Komisaruk *et al.*, 1972).

Hormones and sensory input can be substituted for each other in increasing the probability of eliciting elements of mating behavior. An experiment by Diakow *et al.* (1973) showed that increasing either the amount of sensory input or the level of estrogen increased the probability of eliciting lordosis in female rats. For example, in ovariectomized females given a constant, low dose of estrogen, the probability of lordosis became greater by increasing the amount of stimulation. At a dose level of 30 μg estradiol benzoate, the proportion of females showing lordosis in response to light scratching of the back, flanks, perineum, and clitoris was one out of ten; to pressure applied to these areas it was four out of ten; and to pressure applied to these areas as well as to the cervix it was ten out of ten. When stimulation was held constant, the proportion of females showing good to excellent lordosis in response to injections of oil was two out of ten, and to 0.03 μg and 3.0 μg estradiol, it was four out of ten, and nine out of ten, respectively. In these cases the stimulation was pressure on the back, flanks, perineum, clitoris, and cervix.

The explanation that both sensory input and hormones raise the probability of eliciting mating responses, and that if one of these factors is low, the other can substitute for it, can possibly account for the effects of desensitization of the penis on the mating behavior of male cats. There was a seasonal decline in the tendency to mount after denervation of the penis. According to Aronson and Cooper (1968) sensory input from the penis and androgen were both important for maintaining the tendency of cats to mount; eliminating sensory feedback from the penis lowered this tendency somewhat, but lower androgen levels (reflected by changes in testicular tissue) in the autumn reduced this tendency so far that sex behavior was not shown in some animals at all (Fig. 4).

This concept of the interaction of hormones and sensory input in determining the probability that mating behavior will be shown is similar to Beach's (1944) concept of the contribution of cortical activity and androgen to "sexual excitability" in male rats. He demonstrated that both the presence of the cortex and presence of androgen increased the probabilities of mounting and ejaculating. His interpretation of the data was the following:

> The functions of the cortical mechanism and of the testicular hormone, although they are qualitatively different, exert similar effects upon overt behavior; and they may be regarded as complementary and summative; both contribute to sexual excitability. Since this is true, it is possible, within limits, for a high level of androgen concentration to compensate for a lowered degree of activity in the cortex; and conversely, it is also possible that intensive cortical excitement may temporarily make up for a subnormal androgen level.

VI. Problems in the Analysis of Mating Behavior

A. Difficulties with Interpreting Effects of Deafferentiation or Social Deprivation

It is often difficult to determine what causes changes in behavior after an afferent system has been interrupted or after normal experience has been prevented in other ways. For example, it is hard to know which of multiple cues was used by the animal before deafferentiation. Then too, deprivation may cause changes in the function of the endocrine system and these hormonal changes might mediate the changes in behavior observed afterward. These difficulties are discussed below.

1. Identification of Relevant Cues

Denervation causes loss of sensation in the denervated organs directly, but it may also have broad effects on sexual functioning. Denervation of the vagina and body of the uterus prevents the female cat from responding to intromission, indicating that the vaginal and uterine nerves are used to detect the penis in the vagina through pressure or friction cues. A secondary loss may be involved: this would be the situation if penile contact initiates contractions of the genital tract and these contractions provided feedback stimulation of the female's response to intromission. Masters and Johnson (1966) found that uterine contractions accompany sexual excitement in women.

2. Hormones as Intermediaries of Effects of Deprivation

Any brain lesion, desensitization, or manipulation of a social situation might cause a change in type and amount of hormone secretion which in turn may produce behavioral changes. For example, the effects of olfactory bulb removal on the sexual behavior of female rats have been attributed to the lowering of estrogen levels (Aron *et al.,* 1970). After ablation of the olfactory bulbs, few females mated, but after administering estrogen the percentage that mated increased significantly.

B. Difficulties with Interpreting Effects of "Nonspecific" Stimulation

There is a group of studies relating somatosensory stimulation to mating that is difficult to understand in the context of relating specific stimuli to mating responses. Barfield and Sachs (1968) applied electric shock to the backs of sexually experienced male rats at 30-second intervals in the presence of estrous females. The intensity of the shock was such that the males jumped and squealed slightly at its onset. Seventy-two percent of the mounts with thrusts or intromission followed within 5 seconds of the shock. For the control tests, when these males were not shocked, fewer than 14% of the mount bouts followed the click of the timer for the stimulator. In addition, the interval betweeen the first ejaculation and the following intromission was reduced by 25% in the males when they were being shocked.

Electric shock has also facilitated the appearance of mating behavior in sexually inexperienced male rats and enhanced their tendency to mount other males or a stuffed toy. Shocks increased exploration and investigation of objects in the test cage (food, pieces of wood, water bottle) when females were not present (Caggiula and Eibergen, 1969).

There is no evidence that a male rat must normally be touched on the back or tail before he initiates mounting. So electric shock to the back or tail cannot be interpreted easily as substituting for specific, normally used sensory input which initiates mounting. The results of the above experiments were interpreted to mean that the shock facilitated mounting by increasing sexual arousal. What this means physiologically is hard to determine, but one might guess about involvement of the "reticular activating system" or the limbic system.

The male rat has quiescent periods between intromissions. It is necessary for him to become active again, in the sense that he usually must find the female and walk to her in order to mount her. This may mean that some neural inhibitory process that is caused by intromission or vigorous mounting must wear off before the male can become appropriately active. It could be that the shock, which is presumed to be mildly painful, and was shown to initiate locomotion in both of the experiments cited, overcomes this inhibition and makes the male capable of being more attentive to the cues provided by the female.

Nonspecific stimulation which facilitates mating has also been provided by handling of the male after intromission (Larsson, 1963). This stimulation of the male approximately twice a minute after postintromission licking increased the speed of mating by older male rats. After such handling, males took less time than normal to ejaculate, had shorter intervals between intromissions, and began to intromit sooner after their first ejaculation of the series.

C. DIFFICULTIES IN CHOOSING MEASURES OF RESPONSE

1. Defining a Receptive Female

Experimentalists who deal with the phenomenon of sexual behavior often describe females as being receptive. In the sense that it is used, the term receptivity probably refers to an existing motivational state which predisposes a female to initiate sexual activity or to respond so that mating can occur. Receptivity is measured differently in different species; for example, a female monkey is thought to be receptive if she shows many sexual invitations, such as the presentation posture in which she displays her rear end to the male, and does not refuse the male's attempts to mount (Michael, 1971). A rat is thought to be receptive if she shows the lordosis posture (Ball, 1937), which is an extreme dorsoflexion of the vertebral column. It is the posture or behavior of the female that is actually observed, since it is impossible to observe the motivational state itself.

It is inadvisable to consider the term receptivity as reflecting an existing, unitary motivational state, because the behaviors that are used to measure receptivity may not always vary together. As an example, Hardy and DeBold (1971) have shown differential effects of mounting on two behaviors that are usually considered measures of receptivity in rats: Pretest mounting decreased the probability of rejection behavior (rolling over), but also decreased the probability of lordosis. These effects appear contradictory if one considers failure to reject the male and tendency to show lordosis as two aspects of a unitary phenomenon, receptivity. The implication of this finding is that neither physiologists nor psychologists can study receptivity as a single phenomenon. It is unlikely that there is a unitary neural mechanism underlying receptivity composed of all the neural events which make it possible for a female to mate with a male.

2. *Experimental Manipulation of an Individual Affects the Behavior of Its Mating Partner*

The mating pattern results from interaction and coordination between two partners. Because of this, a behavioral change in one is likely to be reflected in the mating behavior of the other. For example, denervation of the vagina and body of the uterus of cats prevented the females from responding to intromission; but because the females did not respond, the males did not dismount after ejaculation. This means that female cats normally cue the male's dismount (Diakow, 1971a). Likewise, lesions in the sensory area of the thalamus which prevented lordosis in female rats lowered the probability of intromission and ejaculation by the males with whom the females were paired (Diakow, 1971b). It is advisable to study behaviors shown by the male as well as the female even when the experimental manipulation, lesion, social isolation, etc., is applied to only one of them. This will enable us to obtain a full understanding of the factors underlying mating behavior.

VII. SUMMARY

The interaction of a mating pair is governed both by the experiences of the pair previous to the sexual encounter and by their stimulation of each other during the sexual encounter. Specific sensory systems are utilized in the normal manifestation of the mating pattern.

Prepuberal rearing conditions can affect adult mating performance favorably or adversely. The latter occurs when social responses inappropriate for mating develop; for example, the predominate response to another animal of a male rat raised in isolation might be to climb over it from the side, rather than to mount it from the rear.

Mating performance also changes with repeated mating experience. Sexually experienced males ejaculate more quickly, need less preejaculatory stimulation derived from intromission, and have better orientation than sexually naïve males. Females initially tend to become more receptive with repeated matings.

During a sexual encounter, the feedback obtained from performing one portion of the mating sequence affects the performance of subsequent events in copulation. Mounts without intromission shorten the time and reduce the number of intromissions required for ejaculation. Shortening the intervals between intromissions reduces the number of mounts and intromissions to ejaculation and increases the length of each intromission. In addition, this decreases the time necessary before the resumption of copulatory activity after ejaculation. The length of the period of sexual inactivity which follows ejaculation affects the performance of the next copulatory series. When this postejaculatory interval is longer than normal, the interval between intromissions and, consequently, the latency to the next ejaculation is shorter, but a larger number of intromissions are required for this second ejaculation.

During a sexual encounter, the responses of the mating partner contribute to the regulation of an animal's sexual behavior. Females, especially those in heat, approach and remain with males and may prefer certain individual males; but females may actively avoid encounters with males for a while after sexual contact. Unfamiliar females, or females which have been unmated for some time will rearouse males that have mated to satiation. If several females are available to a male simultaneously, he will, however, take longer to ejaculate than when only one female is present.

Specific sensory information guiding copulatory performance in the female rat has been identified to a considerable extent. Touching the back, flank, and perineum causes the female to assume a mating posture, stimulation of the genitalia potentiates holding of the mating posture and initiates the response to intromission. In the male, penile stimulation is necessary for successful ejaculation and, over a long time period, for maintaining the tendency to initiate contact with females. Discrimination of appropriate sexual partners and sexual preferences are based on olfactory cues. Blindness shortens intervals between ejaculations.

Studying the stimulus–response relationships in mating is valuable for understanding the physiological basis of sexual behavior. Knowledge of these relationships has contributed to understanding the relevance of the lateral reticular nucleus of the medulla, the sensory thalamus, the midbrain central gray, and the spinal cord for mating behavior and for understanding the coordination of endocrine and behavioral factors underlying mating behavior. As an example, stimulation obtained in the course of mating triggers both the behavioral afterreaction and hormonal responses to intromission in the female cat. In addition, careful stimulus–response analysis has revealed that both sensory input and gonadal hormones raise the probability of eliciting mating behavior, and if one of these factors is low the other will compensate for it.

Acknowledgments

I should like to acknowledge both the critical reading of an early draft of this manuscript by Lester R. Aronson, Madeline L. Cooper, J. Wayne Lazar, and Daniel S. Lehrman, and the thorough editing of a later draft by Jay S. Rosenblatt. I am also indebted to Kathleen Smith, Charles Malsbury, and Michael Montgomery of Rockefeller University, and to the Reproductive Research Information Service, Ltd., Cambridge, England, for bibliographic assistance; and to Gabriele Zummer and Anne Poston for typing the manuscript. This manuscript was written while I was supported by Public Health Service Grant HD-05751, and a grant from the Rockefeller Foundation.

References

Abrahams, V. C., and Teare, J. L. 1969. Peripheral pathways and properties of uterine afferents in the cat. *Can. J. Physiol. Pharmacol.* **47**, 576–577.

Adler, N. T. 1969. Effects of the male's copulatory behavior on successful pregnancy of the female rat. *J. Comp. Physiol. Psychol.* **69**, 613–622.

Adler, N. T., and Bell, D. 1969. Constant estrus in rats: vaginal, reflexive and behavioral changes. *Physiol. Behav.* **4**, 151–153.

Adler, N., and Bermant, G. 1966. Sexual behavior of male rats: effects of reduced sensory feedback. *J. Comp. Physiol. Psychol.* **61**, 240–243.

Aron, C., Roos, J., and Asch, G. 1970. Effect of removal of the olfactory bulbs on mating behavior and ovulation in the rat. *Neuroendocrinology* **6**, 109–117.

Aronson, L. R. 1965. Environmental stimuli altering the physiological condition of the individual among lower vertebrates. *In* "Sex and Behavior" (F. A. Beach, ed.), pp. 290–318. Wiley, New York.

Aronson, L. R., and Cooper, M. 1963. Genital denervation and sexual behavior in male cats. *Proc. Int. Congr. Zool., 16th, Washington, D.C.* **2**, 24.

Aronson, L. R., and Cooper, M. 1966. Seasonal variation in mating behavior in cats after desensitization of glans penis. *Science* **152**, 226–230.

Aronson, L. R., and Cooper, M. 1968. Desensitization of the glans penis and sexual behavior in cats. *In* "Perspectives in Reproduction and Sexual Behavior" (M. Diamond, ed.), pp. 51–82. Indiana Univ. Press, Bloomington.

Arvidsson, T., and Larsson, K. 1967. Seminal discharge and mating behavior in the male rat. *Physiol. Behav.* **2**, 341–343.

Bacq, Z. 1931a. The effect of sympathectomy on sexual functions, lactation and maternal behavior of the albino rat. *Amer. J. Physiol.* **99**, 444–453.

Bacq, Z. 1931b. Impotence of the male rodent after sympathetic denervation of the genital organs. *Amer. J. Physiol.* **96**, 321–330.

Ball, J. 1934a. Sex behavior of the rat after removal of the uterus and vagina. *J. Comp. Psychol.* **18**, 419–422.

Ball, J. 1934b. Normal sex behavior in the rat after total extirpation of the vasa deferentia. *Anat. Rec.* **58**, 49. (Abstr.)

Ball, J. 1937. A test for measuring sexual excitability in the female rat. *Comp. Psychol. Monogr.* **14**, 1–37.

Bard, P. 1935. The effects of denervation of the genitalia on the oestral behavior of cats. *Amer. J. Physiol.* **113**, 5.

Bard, P. 1939. Central nervous mechanisms for emotional behavior patterns in animals. *Res. Publ., Ass. Nerv. Ment. Dis.* **19**, 190–218.

Bard, P. 1940. The hypothalamus and sexual behavior. *Res. Publ., Ass. Nerv. Ment. Dis.* **20**, 551–579.

Barfield, R. J., and Geyer, L. A. 1972. Sexual behavior: ultrasonic postejaculatory song of the male rat. *Science* **176**, 1349–1350.

Barfield, R. J., and Sachs, B. D. 1968. Sexual behavior: stimulation by painful electrical shock to skin in male rats. *Science* **161**, 392–395.

Beach, F. A. 1942a. Analysis of the stimuli adequate to elicit mating behavior in the sexually inexperienced male rat. *J. Comp. Psychol.* **33**, 163–207.

Beach, F. A. 1942b. Comparison of copulatory behavior in male rats raised in isolation, cohabitation and segregation. *J. Gen. Psychol.* **60**, 121–136.

Beach, F. A. 1944. Relative effects of androgen upon the mating behavior of male rats subjected to forebrain injury or castration. *J. Exp. Zool.* **97**, 249–295.

Beach, F. A. 1945. Hormonal induction of mating responses in a rat with congenital absence of gonadal tissue. *Anat. Rec.* **92**, 289–292.

Beach, F. A. 1947. A review of physiological and psychological studies of sexual behavior in mammals. *Physiol. Rev.* **27**, 240–307.

Beach, F. A. 1951. Instinctive behavior: reproductive activities. *In* "Handbook of Experimental Psychology" (S. S. Stevens, ed.), pp. 387–434. Wiley, New York.

Beach, F. A. 1965. Retrospect and prospect. *In* "Sex and Behavior" (F. A. Beach, ed.), pp. 535–569. Wiley, New York.

Beach, F. A. 1969. Coital behavior in dogs. VII. Effects of sympathectomy in males. *Brain Res.* **15**, 243–245.

Beach, F. A. 1970. Some effects of gonadal hormones on sexual behavior. *In* "The Hypothalamus" (L. Martini, M. Motta, and F. Fraschini, eds.), pp. 617–640. Academic Press, New York.

Beach, F. A., and Holz, A. M. 1946. Mating behavior in male rats castrated at various ages and injected with androgen. *J. Exp. Zool.* **101**, 91–142.

Beach, F. A., and Jordan, L. 1956. Sexual exhaustion and recovery in the male rat. *Quart. J. Exp. Psychol.* **8**, 121–133.

Beach, F. A., and LeBoeuf, B. J. 1967. Coital behavior in dogs. I. Preferential mating in the bitch. *Anim. Behav.* **15**, 546–558.

Beach, F. A., and Levinson, G. 1950. Effects of androgen on the glans penis and mating behavior of castrated male rats. *J. Exp. Zool.* **114**, 159–171.

Beach, F. A., and Rabedeau, R. G. 1959. Sexual exhaustion and recovery in the male hamster. *J. Comp. Physiol. Psychol.* **52**, 56–61.

Beach, F. A., and Whalen, R. 1959a. Effects of intromission without ejaculation upon sexual behavior in male rats. *J. Comp. Physiol. Psychol.* **52**, 476–481.

Beach, F. A., and Whalen, R. 1959b. Effects of ejaculation on sexual behavior in the male rat. *J. Comp. Physiol. Psychol.* **52**, 249–254.

Beach, F. A., and Wilson, J. 1963. Mating behavior in male rats after removal of the seminal vesicles. *Proc. Nat. Acad. Sci. U.S.* **49**(5), 624–626.

Beamer, W., Bermant, G., and Clegg, M. 1969. Copulatory behavior of the ram, *Ovis aries*. II. Factors affecting copulatory satiation. *Anim. Behav.* **17**, 706–711.

Beauchamp, G. K., and Berüter, J. 1973. Source and stability of attractive components in guinea pig (*Cavia porcellus*) urine. *Behav. Biol.* **9**, 43–47.

Bermant, G. 1961. Response latencies of female rats during sexual intercourse. *Science* **133**, 1771–1773.

Bermant, G. 1964. Effects of single and multiple enforced intervals on the sexual behavior of male rats. *J. Comp. Physiol. Psychol.* **57**, 398–403.

Bermant, G. 1965. Rat sexual behavior: photographic analysis of the intromission response. *Psychon. Sci.* **2**, 65–66.

Bermant, G. 1967. Copulation in rats. *Psychol. Today* **1**, 52–61.

Bermant, G., and Taylor, L. 1969. Interactive effects of experience and olfactory bulb lesions in male rat copulation. *Physiol. Behav.* **4**, 13–18.

Bermant, G., and Westbrook, W. H. J. 1966. Peripheral factors in the regulation of sexual contact by female rats. *J. Comp. Physiol. Psychol.* **61,** 244–250.

Bermant, G., Lott, D. F., and Anderson, L. 1968. Temporal characteristics of the Coolidge effect in male rat copulatory behavior. *J. Comp. Physiol. Psychol.* **65,** 447–452.

Bermant, G., Anderson, L., and Parkinson, S. R. 1969. Copulation in rats: relations among intromission duration, frequency, and pacing. *Psychon. Sci.* **17,** 293.

Blandau, R. J., Boling, R. J., and Young, W. C. 1941. The length of heat in the albino rat as determined by the copulatory response. *Anat. Rec.* **79,** 453–463.

Bronson, F. H., and Caroom, D. 1971. Preputial gland of the male mouse: attractant function. *J. Reprod. Fert.* **25,** 279–282.

Brooks, C. McC. 1937. The role of the cerebral cortex and of various sense organs in the excitation and execution of mating activity in the rabbit. *Amer. J. Physiol.* **120,** 544–553.

Brooks, R. J., and Banks, E. M. 1973. Behavioral biology of the collared lemming [*Dicrostonyx groenlandicus* (Traill)]: An analysis of acoustic communication. *Anim. Behav. Monogr.* **6,** Part 1, 1–83.

Bunnell, B. N., and Kimmel, N. E. 1965. Some effects of copulatory experience on postcastration mating behavior in the male hamster. *Psychon. Sci.* **3,** 179–180.

Caggiula, A. R., and Eibergen, R. 1969. Copulation of virgin male rats evoked by painful peripheral stimulation. *J. Comp. Physiol. Psychol.* **69,** 414–419.

Carlson, R. R., and DeFeo, J. J. 1965. Role of the pelvic nerve vs. the abdominal sympathetic nerves in the reproductive function of the female rat. *Endocrinology* **77,** 1014–1022.

Carlsson, S., and Larsson, K. 1962. Intromission frequency and intromission duration in the male rat mating behavior. *Scand. J. Psychol.* **3,** 189–191.

Carlsson, S., and Larsson, K. 1964. Mating in male rats after local anesthetization of the glans penis. *Z. Tierpsychol.* **21,** 854–886.

Carr, W. J., and Caul, W. F. 1962. The effect of castration in the rat upon the discrimination of sex odors. *Anim. Behav.* **10,** 20–27.

Carr, W. J., Loeb, L. S., and Dissinger, M. L. 1965. Responses of rats to sex odors. *J. Comp. Physiol. Psychol.* **59,** 370–377.

Carr, W. J., Loeb, L. S., and Wylie, N. R. 1966. Responses to feminine odors in normal and castrated male rats. *J. Comp. Physiol. Psychol.* **62,** 336–338.

Carr, W. J., Krames, L., and Costanzo, D. J. 1970a. Previous sexual experience and olfactory preference for novel versus original sex partners in rats. *J. Comp. Physiol. Psychol.* **71,** 216–222.

Carr, W. J., Wylie, N. R., and Loeb, L. S. 1970b. Responses of adult and immature rats to sex odors. *J. Comp. Physiol. Psychol.* **72,** 51–59.

Chanel, J., and Vernet-Maury, E. 1963. Reactions de discrimination olfactive et crise andriogene chez la souris. *C. R. Soc. Biol.* **157,** 1020–1024.

Cherney, E. F., and Bermant, G. 1970. Role of stimulus female novelty in the rearousal of copulation in male laboratory rats (*Rattus norvegicus*). *Anim. Behav.* **18,** 567–574.

Cooper, K., Cooper, M., and Aronson, L. R. 1964. Physiological and behavioral observations of erection before and after section of the nerve dorsalis penis in the cat. *Amer. Zool.* **2,** Abstr. No. 114.

Cooper, K. 1969. An electrophysiological study of the effects of castration on the afferent system of the glans penis of the cat. Ph.D. Thesis, New York Univ., New York.

Cooper, M., and Aronson, L. R. 1962. Effects of a sensory deprivation on the sexual behavior of experienced adult male cats. *Amer. Zool.* **2**, Abstr. No. 114.

Cooper, M., and Aronson, L. R. 1972. Persistence of high levels of sexual behavior in male cats following ablation of the olfactory bulbs. *Amer. Zool.* **12**, 657.

Dewsbury, D. A. 1967. A quantitative description of the behaviour of rats during copulation. *Behaviour* **29**, 154–178.

Dewsbury, D. A. 1968. Copulatory behavior in rats: changes as satiety is approached. *Psychol. Rep.* **22**, 937–943.

Dewsbury, D. A. 1969. Copulatory behaviour of rats (*Rattus norvegicus*) as a function of prior copulatory experience. *Anim. Behav.* **17**, 217–223.

Dewsbury, D. A., and Bolce, S. K. 1968. Sexual satiety in rats: effects of prolonged post-ejaculatory intervals. *Psychon. Sci.* **13**, 25–26.

Dewsbury, D. A., and Bolce, S. K. 1970. Effects of prolonged post-ejaculatory intervals on copulatory behavior of rats. *J. Comp. Physiol. Psychol.* **72**, 421–425.

Diakow, C. 1969. The effects of genital desensitization on mating behavior and ovulation in the female cat. Ph.D. Thesis, New York Univ., New York.

Diakow, C. 1970. Effects of genital desensitization on the mating pattern of female rats as determined by motion picture analysis. *Amer. Zool.* **10**, 486.

Diakow, C. 1971a. Effects of genital desensitization on mating behavior and ovulation in the female cat. *Physiol. Behav.* **7**, 47–54.

Diakow, C. 1971b. Effects of brain lesions on the mating behavior of rats. *Amer. Zool.* **11**, 617.

Diakow, C. 1974. Motion picture analysis of rat mating behavior. *J. Comp. Physiol. Psychol.*, in press.

Diakow, C., and Aronson, L. R. 1967. The role of genital sensation in the maintenance of sexual behavior in the female cat. *Amer. Zool.* **7**, 799–800.

Diakow, C., and Aronson, L. R. 1968. Effects of genital desensitization on mating behavior and ovulation in the female cat. *Amer. Zool.* **8**, 748.

Diakow, C., Pfaff, D. W., and Komisaruk, B. R. 1973. Sensory and hormonal interactions in eliciting lordosis. *Fed. Proc., Fed. Amer. Soc. Exp. Biol.* **32**, 241.

Donovan, B. T., and Kopriva, P. C. 1965. Effect of removal or stimulation of the olfactory bulbs on the estrous cycle of the guinea pig. *Endocrinology* **77**, 213–217.

Donovan, B. T., and Traczyk, W. 1965. Interruption of the hypogastric and pelvic nerve supply to the uterus and the occurrence of pregnancy in the guinea pig. *J. Endocrinol.* **33**, 335–336.

Doty, R. L., Carter, C. S., and Clemens, L. G. 1971. Olfactory control of sexual behavior in the male and early-androgenized female hamster. *Horm. Behav.* **2**, 325–335.

Emmers, R. 1965. Organization of the first and second somesthetic regions (SI and SII) in the rat thalamus. *J. Comp. Neurol.* **124**, 215–228.

Fee, A. R., and Parkes, A. S. 1930. Studies on ovulation. III. Effect of vaginal anaesthesia on ovulation in the rabbit. *J. Physiol. (London)* **70**, 385–388.

Fisher, A. 1962. Effects of stimulus variation on sexual satiation in the male rat. *J. Comp. Physiol. Psychol.* **55**, 614–620.

Folman, Y., and Drori, D. 1965. Normal and aberrant copulatory behaviour in male rats (*R. norvegicus*) reared in isolation. *Anim. Behav.* **13**, 427–429.

Folman, Y., and Drori, D. 1969. Effects of the frequency of mating on the androgen-sensitive organs and sexual behavior. *Physiol. Behav.* **4**, 1023–1026.

Fowler, H., and Whalen, R. 1961. Variation in incentive stimulus and sexual behavior in the male rat. *J. Comp. Physiol. Psychol.* **54**, 68–71.

Gerall, A. A., and McCrady, R. E. 1970. Receptivity scores of female rats stimulated either manually or by males. *J. Endocrinol.* **46**, 55–59.

Gerall, H. D., Ward, I. L., and Gerall, A. A. 1967. Disruption of the male rat's sexual behaviour induced by social isolation. *Anim. Behav.* **15**, 54–58.

Greene, E. C. 1963. "Anatomy of the Rat." Hafner, New York.

Gruendel, A., and Arnold, N. J. 1969. Effects of early social deprivation on reproductive behavior of male rats. *J. Comp. Physiol. Psychol.* **67**, 123–128.

Grunt, J., and Young, W. C. 1952. Psychological modification of fatigue following orgasm (ejaculation) in the male guinea pig. *J. Comp. Physiol. Psychol.* **45**, 508–510.

Hard, E., and Larsson, K. 1968a. Dependence of adult mating behavior in male rats on the presence of littermates in infancy. *Brain, Behav., Evol.* **1**, 405–419.

Hard, E., and Larsson, K. 1968b. Effects of mounts without intromission upon sexual behaviour in male rats. *Anim. Behav.* **16**, 538–540.

Hard, E., and Larsson, K. 1968c. Visual stimulation and mating behavior in male rats. *J. Comp. Physiol. Psychol.* **66**, 805–807.

Hardy, D. F., and DeBold, J. F. 1971. Effects of mounts without intromission upon the behavior of female rats during the onset of estrogen-induced heat. *Physiol. Behav.* **7**, 643–645.

Hardy, D. F., and DeBold, J. F. 1972. Effects of coital stimulation upon behavior of the female rat. *J. Comp. Physiol. Psychol.* **78**, 400–408.

Harlow, H. F. 1965. Sexual behavior in the rhesus monkey. *In* "Sex and Behavior" (F. A. Beach, ed.), pp. 234–265. Wiley, New York.

Hart, B. L., and Haugen, C. M. 1971a. Prevention of genital grooming in mating behaviour of male rats (*Rattus norvegicus*). *Anim. Behav.* **19**, 230–232.

Hart, B. L., and Haugen, C. M. 1971b. Scent marking and sexual behavior maintained in anosmic male dogs. *Commun. Behav. Biol. A* **6**, 131–135.

Heimer, L., and Larsson, K. 1967. Mating behavior of male rats after olfactory bulb lesions. *Physiol. Behav.* **2**, 207–209.

Herbert, J. 1967. Neural and endocrine stimuli from the female and the sexual behavior of the male rhesus monkey. *Acta Endocrinol. (Copenhagen), Suppl.* **119**, 47.

Hsiao, S. 1965. The effect of female variation on sexual satiation in the male rat. *J. Comp. Physiol. Psychol.* **60**, 467–469.

Johnson, W. A., and Tiefer, L. 1972. Sexual preferences in neonatally castrated male golden hamsters. *Physiol. Behav.* **9**, 213–217.

Kagan, J., and Beach, F. A. 1953. Effects of early experience on mating behavior in male rats. *J. Comp. Physiol. Psychol.* **46**, 204–208.

Kaufman, R. S. 1953. Effects of preventing intromission upon sexual behavior of rats. *J. Comp. Physiol. Psychol.* **46**, 209–211.

Kollar, E. 1953. Reproduction in the female rat after pelvic nerve neurectomy. *Anat. Rec.* **115**, 641–658.

Komisaruk, B. R. 1972. Induction of lordosis in ovariectomized rats by stimulation of the vaginal cervix: hormonal and neural interrelationships. *In* "Steroid Hormones and Brain Function" (C. H. Sawyer and R. A. Gorski, eds.), UCLA Forum Ser., pp. 127–141. Univ. of California Press, Los Angeles.

Komisaruk, B. R., and Diakow, C. 1973. Lordosis reflex intensity in rats in relation to the estrous cycle, ovariectomy, estrogen administration, and mating behavior.

Endocrinology **93**, 548–557.

Komisaruk, B. R., and Larsson, K. 1971. Suppression of a spinal and a cranial nerve reflex in probing the vaginal cervix in rats. *Brain Res.* **35**, 231–235.

Komisaruk, B. R., Adler, N. T., and Hutchison, J. 1972. Genital sensory field: enlargement by estrogen treatment in female rats. *Science* **178**, 1295–1298.

Krames, L. 1970. Responses of female rats to the individual body odors of male rats. *Psychon. Sci.* **20**, 274–275.

Krehbiel, R. 1948. Reproduction in the rat after cervical bypassing and after cerviotomy. *Anat. Rec.* **101**, 299–318.

Labate, J. S. 1940. Influence of uterine and ovarian nerves on lactation. *Endocrinology* **27**, 342–344.

Larsson, K. 1958. Aftereffects of copulatory activity of the male rat: I. *J. Comp. Physiol. Psychol.* **51**, 325–327.

Larsson, K. 1959a. Experience and maturation in the development of sexual behaviour in the male puberty rat. *Behaviour* **14**, 101–107.

Larsson, K. 1959b. Effects of prolonged postejaculatory intervals in the mating behaviour of the male rat. *Z. Tierpsychol.* **16**, 628–632.

Larsson, K. 1959c. The effect of restraint upon copulatory behaviour in the rat. *Anim. Behav.* **7**, 23–25.

Larsson, K. 1960. Excitatory effects of intromission in mating behaviour of the male rat. *Behaviour* **16**, 66–73.

Larsson, K. 1961a. The importance of time for the intromission frequency in the male rat mating behaviour. *Scand. J. Psychol.* **2**, 149–152.

Larsson, K. 1961b. Duration of facilitatory effects of ejaculation on sexual behavior in the male rat. *J. Comp. Physiol. Psychol.* **54**, 63–67.

Larsson, K. 1963. Non-specific stimulation and sexual behaviour in the male rat. *Behaviour* **20**, 110–114.

Larsson, K. 1969. Failure of gonadal and gonadotrophic hormones to compensate for an impaired sexual function in anosmic male rats. *Physiol. Behav.* **4**, 733–737.

Larsson, K. 1971. Impaired mating performances in male rats after anosmia induced peripherally or centrally. *Brain, Behav., Evol.* **4**, 463–471.

Larsson, K., and Södersten, P. 1973. Mating in male rats after section of the dorsal penile nerve. *Physiol. Behav.* **10**, 567–571.

Larsson, K., and Swedin, G. 1971. The sexual behavior of male rats after bilateral section of the hypogastric nerve and removal of the accessory genital glands. *Physiol. Behav.* **6**, 251–253.

Lashley, K. S. 1938. Experimental analysis of instinctive behavior. *Psychol. Rev.* **45**, 445–471.

LeBoeuf, B. J. 1967. Interindividual associations in dogs. *Behaviour* **29**, 268–295.

Lehrman, D. S. 1962. Interaction of hormonal and experiential influences on development of behavior. *In* "Roots of Behavior" (E. L. Bliss, ed.), pp. 142–156. Harper, New York.

Leonard, C. 1972. Effects of neonatal (day 10) olfactory bulb lesions on social behavior of female golden hamsters *Mesocricetus auratus*. *J. Comp. Physiol. Psychol.* **80**, 208–215.

McGill, T. E. 1962. Reduction in "headmounts" in the sexual behaviour of the mouse as a function of experience. *Psychol. Rep.* **10**, 284.

Malsbury, C. W., Kelley, D. B., and Pfaff, D. W. 1973. Responses of single units in the dorsal midbrain to somatosensory stimulation in female rats. *Progr. Endocrinol., Proc. Int. Congr. Endocrinol., 4th, Washington, D.C.* pp. 229–333.

Mason, W. A. 1960. The effects of social restriction on the behavior of rhesus monkeys: I. Free social behavior. *J. Comp. Physiol. Psychol.* **53,** 582–589.

Masters, W. H., and· Johnson, V. E. 1966. "Human Sexual Response." Little, Brown, Boston, Massachusetts.

Meyerson, B. J., and Lindstrom, L. 1971. Sexual motivation in the estrogen treated ovariectomized rat. *Proc. Int. Congr. Horm. Steroids, 3rd, Hamburg, 1970* pp. 731–737.

Michael, R. P. 1971. Neuroendocrine factors regulating primate behavior. *In* "Frontiers in Neuroendocrinology" (L. Martini and W. F. Ganong, eds.), pp. 359–398. Oxford Univ. Press, London and New York.

Michael, R. P., and Keverne, E. B. 1968. Pheromones in the communication of sexual status in primates. *Nature (London)* **218,** 746–749.

Michael, R. P., and Keverne, E. B. 1970. A male sex-attractant pheromone in *Rhesus* monkey vaginal secretions. *J. Endocrinol.* **46,** xx–xxi. (Abstr.)

Michael, R. P., and Saayman, G. 1967. Sexual performance and the timing of ejaculation in male rhesus monkeys (*M. mulatta*). *J. Comp. Physiol. Psychol.* **64,** 213–218.

Michael, R. P., Keverne, E. B., and Bonsall, R. W. 1971. Pheromone: isolation of male sex attractants from a female primate. *Science* **172,** 964–966.

Missakian, E. A. 1969. Reproductive behavior of socially deprived male rhesus monkeys (*Macacca mulatta*). *J. Comp. Physiol. Psychol.* **69,** 403–407.

Mosig, D. W., and Dewsbury, D. A. 1970. The behavior of rats during copulation as a function of prior copulatory experience. *Psychon. Sci.* **21,** 141–143.

Moss, R. L. 1971. Modification of copulatory behavior in the female rat following olfactory bulb removal. *J. Comp. Physiol. Psychol.* **74,** 374–382.

Murphy, M., and Schneider, G. 1970. Olfactory bulb removal eliminates mating behavior in the male golden hamster. *Science* **167,** 302–304.

Nissen, H. W. 1929. The effects of gonadectomy, vasectomy, and injection of placental and orchic extracts on the sex behavior of the white rat. *Genet. Psychol. Monogr.* **5,** 451–457.

Orsulak, P. J., and Gawienowski, A. M. 1972. Olfactory preferences for the rat preputial gland. *Biol. Reprod.* **6,** 219–223.

Pauker, R. S. 1948. The effects of removing seminal vesicles, prostate, and testes on the mating behavior of the golden hamster (*Cricetus auratus*). *J. Comp. Physiol. Psychol.* **41,** 252–257.

Peirce, J. T., and Nuttall, R. L. 1961a. Duration of sexual contacts in the rat. *J. Comp. Physiol. Psychol.* **54,** 585–587.

Peirce, J. T., and Nuttall, R. L. 1961b. Self-paced sexual behavior in the female rat. *J. Comp. Physiol. Psychol.* **54,** 310–313.

Pfaff, D., and Gregory, E. 1971. Olfactory coding in olfactory bulb and medial forebrain bundle of normal and castrated male rats. *J. Neurophysiol.* **34,** 208–216.

Pfaff, D., and Pfaffmann, C. 1969a. Behavioral and electrophysiological responses of male rats to female rat urine odors. *In* "Olfaction and Taste" (C. Pfaffmann, ed.), pp. 258–267. Rockefeller Univ. Press, New York.

Pfaff, D. W., and Pfaffmann, C. 1969b. Olfactory and hormonal influences on the basal forebrain of the male rat. *Brain Res.* **15,** 137–156.

Pfaff, D. W., Lewis, C., Diakow, C., and Keiner, M. 1973. Neurophysiological analysis of mating behavior responses as hormone-sensitive reflexes. *In* "Progress in Physiological Psychology" (E. Stellar and J. Sprague, eds.), Vol. 5, pp. 253–297. Academic Press, New York.

Rabedeau, R. G., and Whalen, R. 1959. Effects of copulatory experience on mating behavior in the male rat. *J. Comp. Physiol. Psychol.* **52**, 482–484.

Root, W. S., and Bard, P. 1937. Erection in the cat following removal of the lumbo-sacral segments. *Amer. J. Physiol.* **119**, 392–393.

Root, W. S., and Bard, P. 1947. The mediation of feline erection through sympathetic pathways and some remarks on sexual behavior after deafferentiation of the genitalia. *Amer. J. Physiol.* **151**, 80–90.

Rose, J. D., and Sutin, J. 1971. Responses of brainstem neurons to vaginal stimulation in estrous and anestrous cats. *Anat. Rec.* **169**, 414.

Rose, J. D., and Sutin, J. 1973. Responses of single units in the medulla to genital stimulation in estrous and anestrous cats. *Brain Res.* **50**, 87–99.

Rosenblatt, J. S. 1965. Effects of experience on sexual behavior in male cats. *In* "Sex and Behavior" (F. A. Beach, ed.), pp. 416–439. Wiley, New York.

Rosenblatt, J. S., and Aronson, L. R. 1958. The decline of sexual behaviour in male cats after castration with special reference to the role of prior sexual experience. *Behaviour* **12**, 285–338.

Rowe, F. A., and Edwards, D. A. 1972. Olfactory bulb removal: influences on the mating behavior of male mice. *Physiol. Behav.* **8**, 37–41.

Rowe, F. A., and Smith, W. E. 1972. Effects of peripherally induced anosmia on mating behavior of male mice. *Psychon. Sci.* **27**, 33–34.

Sachs, B. D., and Barfield, R. J. 1970. Temporal patterning of sexual behavior in the male rat. *J. Comp. Physiol. Psychol.* **73**, 359–364.

Schein, M. W., and Hale, E. B. 1965. Stimuli eliciting sexual behavior. *In* "Sex and Behavior" (F. A. Beach, ed.), pp. 440–482. Wiley, New York.

Scott, J. W., and Pfaff, D. W. 1970. Behavioral and electrophysiological responses of female mice to male urine odors. *Physiol. Behav.* **5**, 407–411.

Sewell, G. D. 1967. Ultrasound in adult rodents. *Nature (London)* **215**, 512.

Steinach, E. 1894. Untersuchungen zur vergleichenden Physiologie der männlichen Geschlechtsorgane insbesondere der accessorischen Geschlechtsdrusen. *Arch. Gesamte Physiol. Menschen Tiere* **56**, 304–338.

Stern, J. J. 1970. Responses of male rats to sex odors. *Physiol. Behav.* **5**, 519–524.

Stone, C. 1922. The congenital sexual behavior of the young male albino rat. *J. Comp. Psychol.* **2**, 95–153.

Stone, C. 1923. Further study of sensory functions in the activation of sexual behavior in the young male albino rat. *J. Comp. Psychol.* **3**, 469–473.

Stone, C. 1925. The effects of cerebral destruction on the sexual behavior of male rabbits. I. The olfactory bulbs. *Amer. J. Physiol.* **71**, 430–435.

Stone, C. 1927. The retention of copulatory ability in male rats following castration. *J. Comp. Psychol.* **7**, 369–387.

Stone, C., and Ferguson, L. W. 1940. Temporal relationships in the copulatory acts of adult male rats. *J. Comp. Psychol.* **30**, 419–433.

Thompson, M. L., and Edwards, D. A. 1972. Olfactory bulb ablation and hormonally induced mating in spayed female mice. *Physiol. Behav.* **8**, 1141–1146.

Tiefer, L. 1969. Copulatory behavior of male *Rattus norvegicus* in a multiple-female exhaustion test. *Anim. Behav.* **17**, 718–721.

Valenstein, E. S., Riss, W., and Young, W. C. 1955. Experiential and genetic factors in the organization of sexual behavior in male guinea pigs. *J. Comp. Physiol. Psychol.* **48**, 397–403.

Whalen, R. 1963a. The initiation of mating in naive female cats. *Anim. Behav.* **11**, 461–463.

Whalen, R. 1963b. Sexual behaviour of cats. *Behaviour* **20,** 321–342.

Wilson, J., Kuehn, R., and Beach, F. A. 1963. Modifications in the sexual behavior of male rats produced by changing the stimulus female. *J. Comp. Physiol. Psychol.* **56,** 636–644.

Young, W. C. 1961. The hormones and mating behavior. *In* "Sex and Internal Secretions" (W. C. Young, ed.), Vol. 2, pp. 1173–1239. Williams & Wilkins, Baltimore, Maryland.

Author Index

Numbers in italics refer to the pages on which the complete references are listed.

Subject Index

A

Acoustic receptor, spike response of, 29–32

Adaptedness, of neuronal mechanisms, 13–14

Afferent pathways, *see also* Deafferentation
mating and,
auditory, 251
olfactory, 245–250
tactile, 238–245
visual, 250–251

Alternating activities, in behavior sequences, 222–224

Anosmia, attempts to induce, 245–247

Atmospheric factor, homing and, 80–83

Audition, *see* Hearing

B

Behavior sequences, time-sharing and, 205–218
displacement activities and, 218–220
functional considerations, 220–224

Birds, homing in, *see* Homing

Brain,
embryonic behavior and, 145–150
mating and, 251–252

C

Chick embryo, *see* Embryonic behavior

Competition, time-sharing and, 201–205

D

Deafferentation, interpretation of effects on mating, 255–256

Deprivation,
sensory, embryogenesis of behavior and, 157–161
social, effects on mating, 255–256

Disinhibition, time-sharing and, 201–205

Displacement activities, 220–222
as components of behavior sequences, 218–220

Distance, homing and, 101–103

E

Ejaculation, mating pattern and, 234

Embryonic behavior, 133–135
brain influences on, 146–150
chronology of, 135–141
electrophysiological and neurochemical correlates of, 150–155
motor primacy and endogenously produced motor behavior, 155–157
neurogenic *vs.* myogenic basis of, 145–146
ontogenetic organization of motor behavior, 141–145
sensory deprivation and augmentation and, 157–161

Evasive behavior, palp and pilifer and, 37–38

F

Flight, spontaneous activity and, 26–27

G

Genitals,
grooming of, 235
stimulation of, 239–245

H

Hawkmoths, evasive behavior in, 37–38

Hearing,
mating and, 251
in noctuid moths, 29–37

Homing, 47–48, 116–118
capabilities for, 48–50
experimental techniques for, 109–116
inertial guidance in, 77–79
landmarks and, 86–92
magnetism and, 70–77
map-and-compass model of, 63–68
map component in, 92
release site and, 92–106
temporal changes and, 106–109
Matthews' hypothesis, 54–63
meteorological cues in, 85–86
nocturnal, 64–65
olfactory cues in, 83–85
orientation types in, 50–53
stars and, 79–80
Wallraff's "atmospheric factor" and,